AF695374

The Identification of the Northern European Woods

The Identification of the Northern European Woods

A guide for archaeologists and conservators

Jon G. Hather

Routledge
Taylor & Francis Group
LONDON AND NEW YORK

First published 2000 by Archetype Publications Ltd

2 Park Square, Milton Park, Abingdon, Oxfordshire OX14 4RN
52 Vanderbilt Avenue, New York, NY 10017

Routledge is an imprint of the Taylor & Francis Group, an informa business

First issued in paperback 2020

ISBN 13: 978-1-873132-47-0 (hbk)
ISBN 13: 978-0-367-60563-6 (pbk)

Typeset by Kate Williams, Abergavenny.

Contents

Part Two: Identification of Northern European Woods

Part Three: Character Tables

Acknowledgements

This book would not have been possible without the help of Sandra Bond who undertook much of the microtomy of the woods included in the volume, Sarah Mason who took many of the photographs and Stuart Laidlaw who developed and printed the photographs. I am grateful to the National Forestry Commission's arboretum at Westonbirt, Wiltshire for providing some of the wood specimens, and to my family who patiently helped me collect others.

PART ONE

Introduction

Introduction

Although wooden artefacts are less commonly preserved on archaeological sites than artefacts made from more durable materials such as stone and ceramics, there is throughout prehistory and later historic periods mounting evidence to suggest that wood was the most widely used resource of the past. Not only did it provide a fuel for heating, light and cooking it also provided the fuel for working metals and firing pots. Wood was the most important material for building and, without question, the most ubiquitous resource for manufacturing domestic utensils, plates, bowls, furniture, agricultural and fishing equipment, all manner of tools, machines for industries, for boats and for land transport. Most of the material culture of the past was wooden and much of it has decayed and disappeared.

Some archaeological sites, however, are sufficiently wet, dry or cold for wood to be preserved. In some cases wood is preserved on such sites to the extent that a past where wood dominated as a material for building, for manufacturing domestic, industrial or agricultural tools and machines is abundantly evident. Charcoal – wood that has become burnt either through intention or accident – is on the other hand less vulnerable to attack by micro-organisms and fungi. Its presence on the vast majority of archaeological sites is testament to the almost universal use of wood as a fuel.

The aim of this book is to provide a methodology for the identification of the woody taxa used to manufacture artefacts recovered from archaeological excavations. The book is divided into three parts: the first part briefly summarises methods; the second part provides the anatomical descriptions of the woody taxa used in artefact manufacture, as structural elements and for fuel in Northern Europe; and the third part provides list of characters for the taxa included in this volume.

'Northern Europe' – geographical and botanical definitions

In certain respects the area under consideration here is naturally defined by both physical boundaries and by floristic communities. It is roughly defined by a line in France east from the Loire valley through to the northern foot hills of the Alps in Germany, cutting across the Carpathian mountains to the north of the Black sea to the Volga hills in southern Russia. The area is delimited to the east by a line from the Volga hills in the south to the Kola peninsula in the north. The area includes the Ireland and Iceland in the west and Scandinavia in the north.

The temperate flora, dominated by deciduous woodlands of oak, beech, ash and hornbeam, occupies much of the area, reaching up as far as southern Sweden and northwest Russia. The northern boreal forest dominated by pine, spruce and birch sweeps across northern Scandinavia into Russia.

The woody flora is not large – some 200–250 species. This volume includes all the tree and larger shrub taxa. The objective is to aid in the identification of wood from archaeological contexts, wood that has been used in the past largely for the manufacture of artefacts and structures or has been used for fuel. The smaller shrubby taxa are rarely used in this respect and therefore have not been included. The major group of woody taxa not included are the Ericaceous shrubs – Erica, Arctostaphyllos, Phyllodoce, Camedaphne and so on – few of which will achieve a basal diameter of over 15 mm.

The following terms have been used with respect to tree and shrub form. A full list of the forms of each of the woody taxa included in this volume is found in Tables 3.19–3.21 in Part 3.

Term	Definition	Example
Large tree	single-stemmed and over 10 m	*Prunus avium*
Medium tree	single-stemmed and between 3 and 10 m	*Malus sylvestris*
Small tree	single-stemmed and up to 3 m	*Euonymus europaeus*
Large shrub	multi-stemmed and 2.5 m+	*Hippophae rhamnoides*
Medium shrub	multi-stemmed and between 1 and 2.5 m	*Suaeda vera*
Climber	climbs by attaching itself to a surface	*Hedera helix*
Liana/scrambler	supports itself by climbing through other plants	*Lonicera periclymenum*

Introduction to the anatomical characters

Wood is a complex biological tissue formed as part of a plant's body to enable both an increase in the plant's size and in its longevity. Identification of wood is achieved through the recognition of a series of anatomical characters which vary between species, genera and families. These are viewed through thin sections of woody tissue – three sections each taken at 90 degrees to each other (see Figure 1.1). The characters that are used to identify wood are therefore seen as two dimensional although it must be remembered that each of these characters is part of a three-dimensional tissue.

The first of the three sections cuts across the trunk or branch of a tree and is known as a cross section or, more commonly, transverse section (TS). The other two sections are at 90 degrees to the TS running down the axis of the trunk or branch. The first of these follows the radius of the trunk or branch and is known as the radial longitudinal section (RLS). The other longitudinal section is at 90 degrees to this and is concentric to the annual rings of growth of a tree. This is the tangential longitudinal section (TLS).

The constituent cells of wood fall into two groups – those that make up the axial system and those that make up the radial system. The axial system is mostly dominant in terms of cell numbers and cell types. Vascular cells within the axial system transport water along the long length of the plant axis. The simpler radial system is made up of rays only and transports water across the radius of the axis (see Figure 1.2).

Figure 1.1 Locating TS, TLS and RLS.

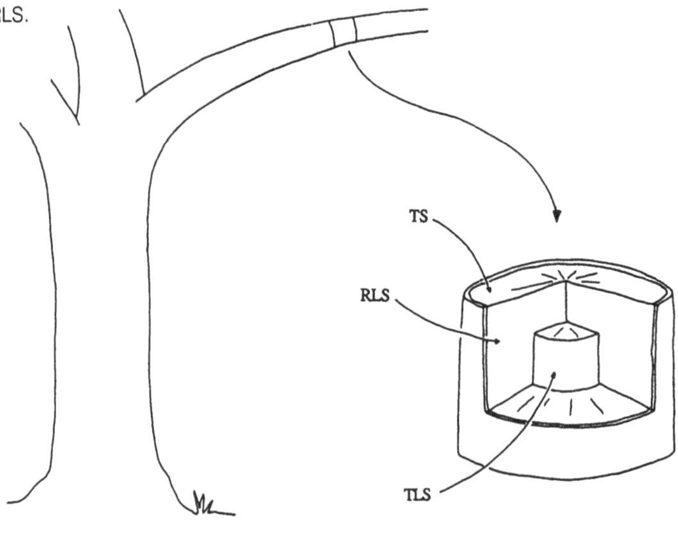

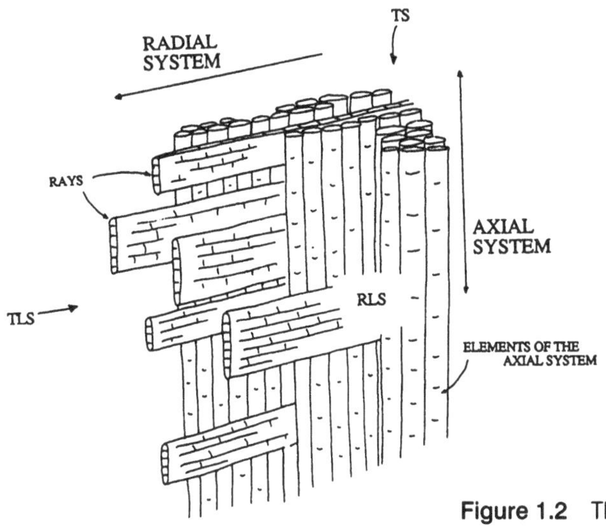

Figure 1.2 The axial and radial systems.

The three sections used to examine the diagnostic characters of wood's anatomy, TS, RLS and TLS, each view the axial system and the radial system from different perspectives so that the same features and the same characters can be seen from different positions. In this way examination of the TS, RLS and TLS can build up into a three-dimensional view of the wood structure.

In this book the woods of northern Europe are divided into eight anatomical groups based upon the organisation of the constituent cells of the axial system seen in TS. Seven of the eight groups are Dicotyledons, the broad-leaved trees and shrubs often referred to as hardwoods, which have a comparatively complex anatomy. The other group, and the first to be dealt with, is the Gymnosperms or the softwoods, which have a relatively simple anatomy.

The wood of the Gymnosperms

The axial system of the Gymnosperms is made up entirely of cells called tracheids. These are long, thin tapering cells that have along their length bordered pits which allow water to pass between them. Bordered pits may be unisereate, i.e. a single row (see Figure 1.3i), or in pairs with thickened bars called crassulae above and below them (see Figure 1.3ii). The tracheids become smaller and thicker walled as growth across a single year proceeds, though this may happen gradually or abruptly.

The rays of the radial system are all unisereate, that is to say that they are only one cell wide. However, they vary in height from one or two cells up to 40 cells or more. Rays may be homocellular or heterocellular. Homocellular rays are composed of procumbent cells only while heterocellular rays have ray tracheids at their margins (the top and bottom) (see Figure 1.4). These tracheids may be smooth-walled or have dentate internal walls.

The pits that allow water to pass from the axial system to the radial system are of considerable diagnostic significance. These pits are located in the ray wall where it abuts against a tracheid adjacent to it. The horizontal walls of the procumbent cells of the ray against the vertical walls of the tracheids results in a series of square or rectangular areas known as the

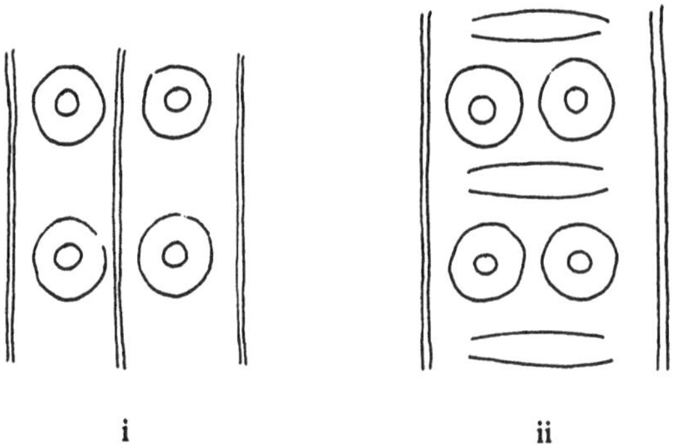

Figure 1.3 Bordered pits (i) unisereate bordered pits (ii) bordered pit pairs with crassulae.

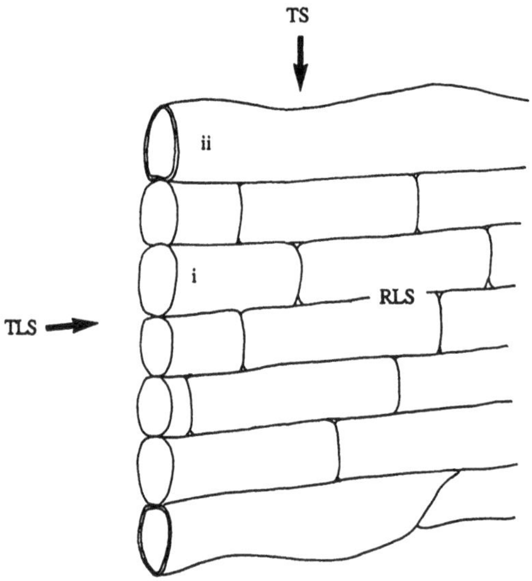

Figure 1.4 Gymnosperm ray (i) procumbent body cells and (ii) marginal ray tracheids (present in some taxa).

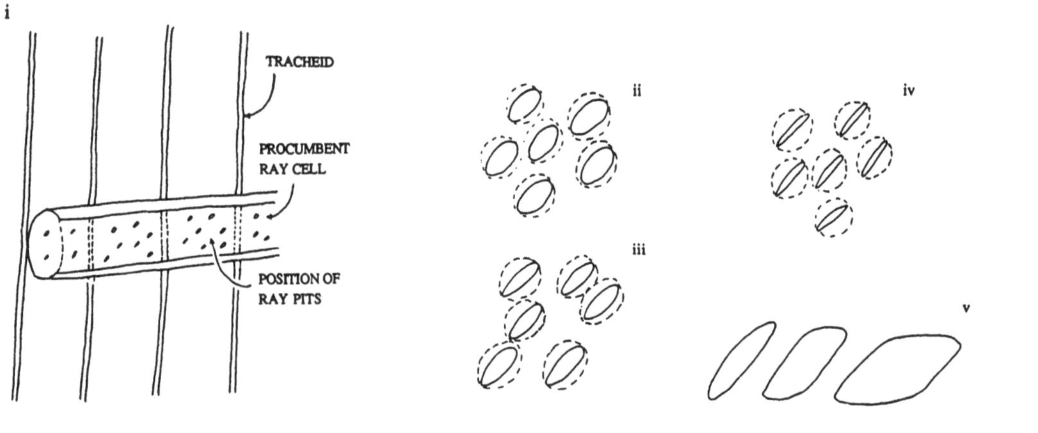

Figure 1.5 Ray pitting. (i) the position of ray pits – 'the cross-field' (ii) taxodioid (iii) cupressoid (iv) piceoid (v) fenestriform (one per cross-field).

cross-field (see Figure 1.5i). The pits, called cross-field or ray pits, vary in size and shape. The largest are fenestriform pits, found in *Pinus sylvestris* and *P. mugo* (see Figure 1.5ii). The three other pit types are much smaller (see Figures 1.5iii, iv and v).

Resin ducts are present in some species in both the axial and the radial systems. These are linear channels bordered by epithelial cells. In the axial system they follow the axis of the branch or trunk and appear in TS as circular holes, usually in the late wood (ie wood laid down towards the end of a year's growth) where tracheids are often thicker walled than in the early wood. In the radial system the resin ducts are embedded within rays. Rays containing resin ducts may be up to three cells wide above and below the duct. The epithelial cells of resin ducts may be thick- or thin-walled.

Parenchyma is present as diffuse solitary cells in the late wood of *Juniperus* species. Here the contents are coloured and these cells are easily seen.

The wood of the Dicotyledons

Both the axial and the radial systems of the Dicotyledons are more complex than those of the Gymnosperms. The axial system is made up of many more cell types that have a variety of functions. The radial system not only has unisereate rays but also rays that can very be very much larger.

The axial system comprises a variety of cells that have a mechanical and vascular function – libriform fibres, fibre tracheids and vascular tracheids – however most of the vascular function of the axial system is undertaken by vessels made up of many individual vessel elements stacked up on top each other. Vessels, known also as pores especially with reference to their appearance in TS, are of great diagnostic significance. In this book their organisation – described in the key to anatomical groups in Part 2 (see pp. 26–7) – is central to identification.

Of the seven anatomical groups described, one, represented by only one species, has vessels associated with phloem within the secondary xylem in the form of bundles. This is a very characteristic organisation and is difficult to confuse with other dicotyledonous woods (see pp. 26 and 44).

Of the remaining six the vessels or pores are arranged in characteristic patterns across individual annual growth rings. There are two distinct patterns:

- in one there is a significant difference between the size of the vessels put down in the early months of growth within a year, which are large, and those put down later which are much smaller. This is described as **ring porous** wood
- in the other the difference in the size of vessels or pores between the early and late growth of a year is not significant, if there is a difference at all. This is described as **diffuse porous** wood.

There is a third category that describes an organisation of vessels or pores that is intermediate between ring porous and diffuse porous wood, known as **semi-ring porous** wood. Most woods that can be semi-ring porous are also often diffuse porous and so this has not been used as a major diagnostic character. All woods that have a tendency towards being semi-ring porous have been included within diffuse porous groups but this character's state is indicated.

Within diffuse porous woods and within the late wood of the ring porous woods the vessels can be arranged in different ways (see Part 2, pp. 26–7). These are solitary vessels, vessels in clusters or radial files and vessels in a dendritic pattern.

Vessels are made up of many separate vessel elements stacked up upon each other. The end walls of the vessel elements have to allow water to pass through and so are open. This opening is called a perforation plate and may be simple (a circular hole) or scalariform (with parallel bars). The perforation plate is angled within the vessel element so that is most visible within the RLS (see Figure 1.6).

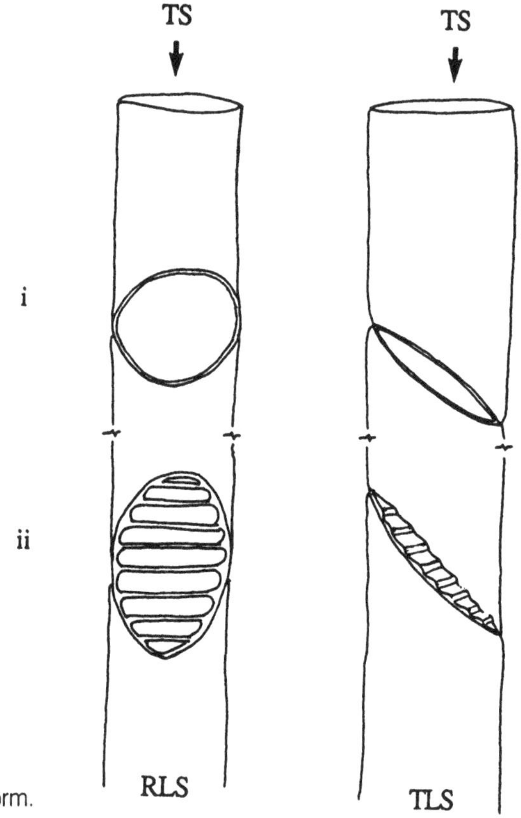

Figure 1.6 Perforation plates (i) simple (ii) scalariform.

Rays may be unisereate and composed of procumbent cells alone. These are the simplest of rays and occur in many taxa. However there are two variations on this simple structure. First, rays may be many cells wide and often with an increase in width there is an increase in height. A number of taxa, such as *Quercus*, have rays that are broad and high enough to be visible to the naked eye. In *Quercus* both unisereate rays and the larger multisereate rays are made up of procumbent cells only. The second variation on this simple ray structure is where there is a combination of procumbent ray cells with square or upright cells (see Figure 1.7). Other rays can be made of upright cells only (see Figure 1.8). Where a ray is made up of a single ray cell type, whether procumbent or upright, the ray is termed homogeneous. Where there are procumbent ray body cells with upright or square marginal cells, the ray is termed heterogeneous. There are three categories of heterogeneous rays:

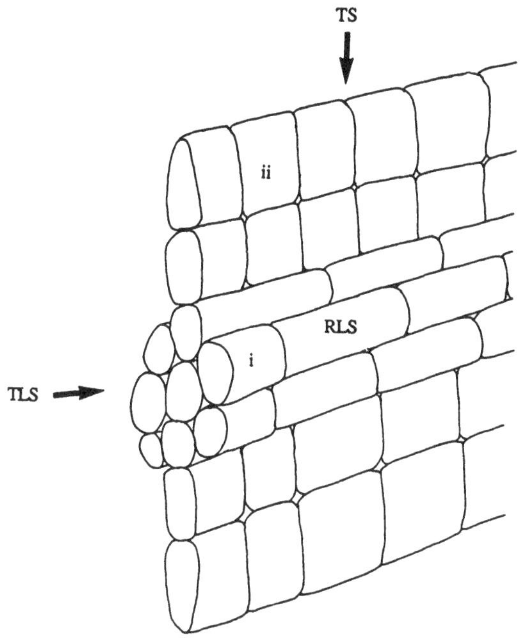

Figure 1.7 Multisereate heterogenous ray with (i) procumbent body cells (ii) upright/square marginal cells.

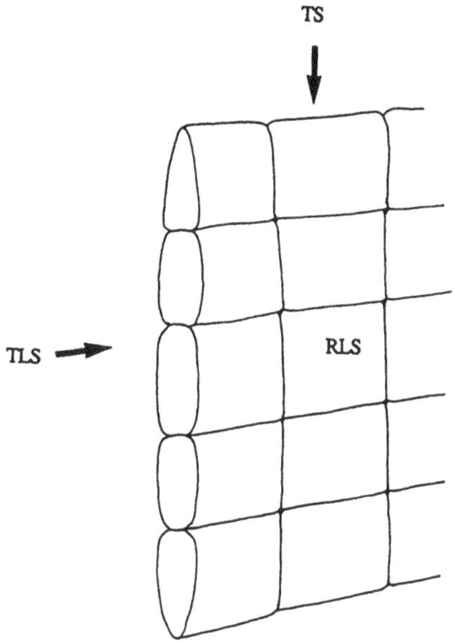

Figure 1.8 Unisereate ray composed of upright cells only.

- heterogeneous type 1 – with many rows of square or upright marginal cells
- heterogeneous type 2 – with two rows of square or upright marginal cells
- heterogeneous type 3 – with one row of square or upright marginal cells.

Aggregate rays occur in the Betulaceae, though are rare in *Betula* itself. An aggregate ray is an aggregation of relatively small rays within the structure of the wood that appear very close together with the absence of vessels between them. Functionally they are similar to large multisereate rays and at very low power magnification they appear alike. Aggregate rays appear in three of the four taxa that fall within the diffuse porous group that has vessels in long radial files. These are *Corylus*, *Alnus* and *Carpinus*.

Sheath cells are square or upright cells that are found sheathing the body of multisereate rays and are most visible in TLS (see Figure 1.9).

Parenchyma is much more common and varied in position in the Dicotyledons than in the Gymnosperms. There are two positions – Apotracheal and Paratracheal. Paratracheal parenchyma is found adjacent to vessels and apotracheal parenchyma is found within the ground tissue. Parenchyma is visible from within all sections but is most obvious in TS. Apotracheal parenchyma is most common and is usually diffuse, comprising solitary cells or small clusters. Some taxa have tangential bands of apotracheal parenchyma. Apotracheal parenchyma is also seen at the end of a year's growth where it is known as terminal parenchyma. Paratracheal parenchyma is less common and often difficult to see, however, *Fraxinus* has very obvious and diagnostic paratracheal parenchyma surrounding the late wood vessels.

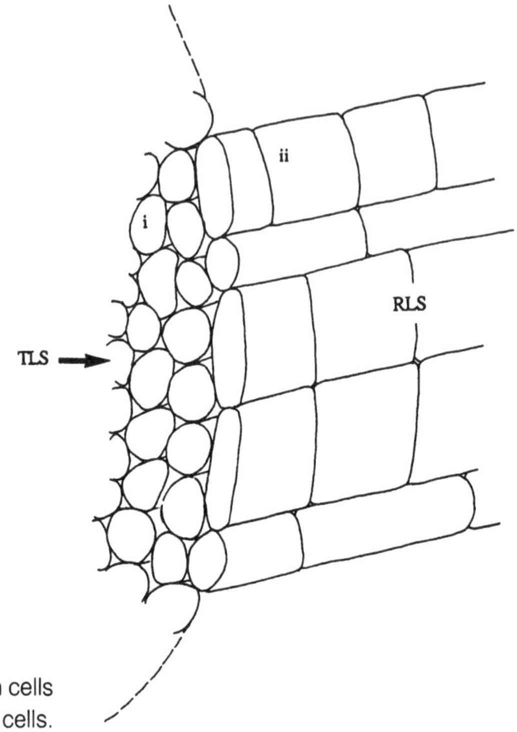

Figure 1.9 Section of multisereate ray with sheath cells (i) sheath cells and (ii) procumbent body cells.

The variability of anatomical characters

The anatomical characters described for the taxa in this volume, and the photographs alongside them, present a snapshot of the anatomy of each. While these basic character states will remain stable for all individuals of each taxon, there will always be a degree of variation visible. This commonly results from external environmental factors and internal growth patterns or may have a genetic basis.

Environmental factors

The growing position of a tree or shrub can determine the amount of growth it achieves on an annual basis. A tree growing in optimum conditions (these will vary from species to species) will put on more growth than one which is waterlogged or has too little water, is growing in a position where it has too little light, not enough nutrients, and so on. Seasonal variation in climate, most notably rainfall, will also cause a tree to put on less growth than it normally might. Variation from the normal pattern of growth caused by these factors is often in the form of reduced annual growth. Sometimes this results from dry summer months and is visible in the form of reduced late wood growth. This is difficult to determine in diffuse porous woods but in ring porous woods the effect is much more noticeable, especially if such summer droughts continue year after year. Slow grown woods resulting from environmental stress also commonly have smaller and thicker walled cells and ray size may be reduced.

Growth patterns

Young wood commonly lacks a range of features that may be present in more mature wood of that taxon. This is mostly due to the small size of annual rings of growth in young wood and results in some characters that are large in mature wood to be much smaller. The size of large rays and early wood vessels and the length of pore files may all be reduced in immature wood. Young wood may also lack well-developed late wood.

The anatomy of damaged wood and wood around knots as well as wood associated with insect, fungal, wind or frost damage may also be different from normal wood. Often vessel frequency is much lower and ground tissues are thickened. The normal linear alignment of tissues may also be affected.

Note
This section, describing the characters used to identify wood, is particular to both the flora under discussion here and to the way in which the method for identification is presented in this volume. For both these reasons the section outlines some characters and character states and leaves others out. I acknowledge it is also brief though it is designed as an introduction sufficient for this guide to identification to be used. Readers wishing to build on this brief section should consult the excellent volume by Jane (1956) which sadly is only available in libraries and second-hand bookshops, or to the more recent and still available Wilson and White (1986). For a detailed account of the features used in hardwood identification readers are referred to Wheeler *et al.* (1989) and for softwood identification to Phillips (1948).

The level of identification of archaeological wood

Most trees and shrubs can be readily identified to the level of species through the morphological characters of their leaves, flowers, branching pattern, twigs, buds and overall size and shape. Identification to the level of species is however not always possible using characters of the wood anatomy. The level of identification varies between genera and families. At the beginning of Part 2 of this volume is a list of the identifiable groups of taxa divided into anatomical groups for ease of identification. These represent the highest level of identification possible – some are species, some are groups of species within a genus, many are genera, some are sub-families or families. Here are some examples:

- Identifiable to the level of family: Leguminosae within which the wood anatomy is as variable within individual species as it is across the family, and the Salicaceae within which the two genera, *Salix* and *Populus* can often not be separated because of the similarity in their anatomy.
- Identifiable to the level of sub-family: Maloideae (Rosaceae) where the genera *Malus*, *Pyrus*, *Sorbus*, *Crategus* cannot be differentiated.
- Identifiable to genus: *Quercus*, *Ulmus*, *Rosa* – the species of these three genera cannot be separated.
- Identifiable to the level of species group: *Acer campestre*, *A. opulus* and *A. monspessulanum* as a group cannot be differentiated from each other but can be differentiated from *Acer platinoides* and *A. pseudoplatanus*.

- Identifiable to the level of species: *Viburnum lantana* and *Viburnum opulus* can be separated as can *Cornus sanguinea* and *Cornus mas*, though with difficulty.

Occasionally two separate genera are difficult to differentiate, for example *Larix* and *Picea*, both Pinaceae, and *Lonicera xylosteum* and *Ligustrum vulgare*. The latter pair are of two different families – Caprifoliaceae and Oleaceae.

Where a wood is identifiable to the level of species this may be because it has characters to differentiate it from other species of the same genus. Often though it may be identifiable to the level of species because it is the only species of that genus present in the flora. If other species of the genus from other regions were present (perhaps as introductions) it may not be possible to separate them. This is true of *Juniperus communis* where it occurs in northern Europe, north of the natural distribution of *Juniperus sabina*, and of *Alnus glutinosa* in Britain where *Alnus incana* is not native. For both *Juniperus* and *Alnus* the two species cannot be separated on anatomical characters.

In many cases separation of species of a genus is possible, but is based on characters that require good preservation of the material (e.g. separating two *Sambucus* species or the two *Cornus* species) or require a large sample so that many individual examples of a particular character can be observed on a single wood section (e.g. separating *Acer* or *Prunus* species on the character of ray size). Both types of characters may be difficult to obtain in archaeological material.

Methods for identification

There are two components to the identification of archaeological wood: anatomical identification – the characters for which have been introduced – and prior to this the process of handling the wood, understanding the direction of the grain, selection of the area for sampling and taking the sample for identification.

For each of the different modes of preservation of archaeological wood, waterlogged, desiccated or charred, the anatomy of the three sections TS, TLS and RLS (see pp. 4–5 and Figure 1.1) has to be examined. The methods for sampling wood for identification are outlined here.

There are many different ways of sectioning wood to produce a thin enough sliver to be able to see the anatomical structure. For many purposes, including making a reference collection, a microtome, either manual or automatic, sledge or rotary will produce multiple large, evenly thick sections from which publication quality photographs can be taken. However, a microtome has two disadvantages for identifying archaeological wood. First, setting up the wood to be sectioned on a microtome takes time, and secondly, the wood to be sectioned must be relatively robust. Since archaeological wood is often anything other than robust and for identification numerous publication quality sections are not necessarily needed, microtomes are rarely used. Hand-cut sections are often less uniformly thick though they are much quicker to take.

Sectioning archaeological wood is best done by hand with half of a double-sided razor blade (the type usually used for shaving). This has a flexibility that single-sided razor blades, which have a protective rigid metal back, and scalpels lack. How you hold the razor blade is a matter of personal preference, however, it is often held most easily with your middle and/or index finger underneath and your thumb on top. This allows very good control of the direction of the cut and the flexing of the edge. Charcoal has a structure that is much less easily sectioned and is better examined under reflected light. Clean surfaces are produced by fracturing, using the pressure of a sharp edge, although for this process the rigidity of a single-sided razor blade or scalpel is required.

When sectioning wood with a razor blade one of the most valuable skills to develop is the recognition of a blunt razor blade. Many beginners continue to use blunted razor blades and blame their lack of experience for the poor sections produced. How long a razor blade will last will depend on the species and nature of preservation of the wood as well as the make of blade. Experiment by trying a new razor blade if you are having difficulty in cutting a section. For sectioning wood the equipment needed is as follows:

- double sided razor blades (snapped in two)
- industrial methylated spirit (IMS)
- pipette
- fine paint brush
- glass microscope slides
- glass cover slips
- absorbent tissues

A note on microscopy

The identification of wood using anatomical characters requires the use of microscopy. There are two forms of optical microscopy that vary in the way in which the object being viewed is lit. Transmitted light microscopy has the light source beneath the subject and the objective lens of the microscope above it. Light is transmitted through the subject and in order to allow this to happen the subject must be very thin, hence the thin sectioning of wood discussed above. Epi-illuminated or reflected light microscopy has its light source above the subject. Light is reflected from the subject into the objective lens and so this form of microscopy is ideal for the examination of objects that cannot be made into thin sections (easily or at all) such as wood charcoal. Scanning electron microscopy is a good substitute for epi-illuminated microscopy where very small and friable fragments of charcoal need to be identified. The increased depth of field, as well as the ease of making a photographic record, outweighs the sometimes time-consuming preparation of samples. However, scanning electron microscopy is not suitable for tackling large assemblages of charcoal where the preparation time would be difficult to justify.

The magnification that may be achieved by an epi-illuminated or transmitted light microscope is determined by multiplying the magnification of the eyepieces by that of the objective lens. Eyepieces are commonly 8×, 10×, 12× or 15×. Objective lenses may be 4×, 6×, 10×, 20×, 40×, 60× or 100×. Very broadly speaking, an overall view of the constituent tissues of a section of wood will be achieved at approximately 40× to 100×. It must be remembered, however, that the diameter of the whole of a 20-year-old stem of a small shrub would easily fit into two or three growth rings of a fast-grown oak. Detailed examination of the tissues of any wood requires 100× to 600× magnification. Using the upper end of this magnification it is possible to count the number of bars of a scalariform perforation plate or to examine the cross-field pits in a Gymnosperm RLS. Higher magnification than this is not needed for the identification of the woods described here but is used in wood identification to examine pores and pits. Lower power reflected light microscopy is used to examine charcoal and to fracture it ready for examination under higher power microscopy. The magnifications most useful here are between 6× and 40×.

Identifying artefactual and non-artefactual wood

Wooden artefacts may be preserved by waterlogging, desiccation, freezing or, very rarely, by charring. Wooden objects from the relatively recent past are also often preserved intact having passed through generations of ownership being carefully looked after. Such artefacts are commonly held in museum collections and may only date to within the past six or seven centuries, however, they can be very informative of the past. Ethnographic collections similarly may date only to the recent past but knowing the identification of wooden components of such artefacts may add valuable information to an understanding of their manufacture and use.

Whichever way an artefact is preserved it is very important to examine it very carefully. Two basic observations have to be made prior to sampling. The first is to determine the direction of the grain and to orient the object with respect to the branch or trunk from which it has been made. This will allow you to locate areas that expose the three planes, TS, TLS and RLS, for sectioning. The second is to examine the object carefully for areas that would be much better for sectioning than others, avoiding damage to the surface where toolmarks, surface pigments and so on may be present.

Determining grain direction

For many objects the direction of grain is obvious, commonly because the wood is ring porous (see pp. 26 and 46–70) or because the object shares its long axis with that of the branch or trunk from which it was converted. This, however, is not always the case and difficulties may arise for several reasons:

- the shape of the object may be rounded or hollowed out
- toolmarks (especially saw marks) can obscure grain direction
- deep staining or paint will obscure grain direction
- if the object has been converted obliquely from the branch or trunk
- for some species grain itself is naturally difficult to see
- when the grain itself does not follow a straight line or is particularly knotty .

A good understanding of the technology used to manufacture artefacts, such as cooperage, turning, joinery and carving, will help to locate grain direction. However, it must not be assumed that simply because wood is usually converted in a particular way for the manufacture of an object that this will always have been so. Mistakes, experimentation and variation in technology are just as likely to have been prevalent in the past as they are today.

Anatomical features visible to the naked eye
One of the most obvious features in many woods are the annual rings. These are seen as concentric circles in transverse section and as linear features in RLS. In TLS the pattern of the annual rings depends on their curvature, width and alignment, but broadly speaking these are linear too. In some woods however annual rings are very faint or may not be visible.

Rays in some taxa, where they are large enough to be seen with the naked eye, may also be helpful in determining the direction of grain. In TS rays are seen as linear features radiating out from the centre of the log from which an object has been converted, crossing the annual rings at 90 degrees. Rays vary in their length but often they will be visible across several annual rings in TS. In RLS rays are often seen as small rectangular and sometimes reflective flecks, and in TLS as short vertical lines. It must be stressed that rays will only be helpful in locating the direction of grain in taxa where such large rays are present, e.g. *Quercus* sp. (oaks) and *Fagus sylvatica* (beech) (see Part 3, Table 3.11, p. 176 for a full list).

The axial system of wood comprises many cells that are linear in form and all aligned parallel to the long axis of a trunk or branch. The long axis of the grain therefore, will have a fibrous nature, though without obvious rays or annual rings it may be difficult to differentiate the TLS and RLS. If the LS can be determined, any section at 90 degrees to it will be a transverse section. Careful examination of the features of the TS located on to the wood being sectioned should indicate where the TLS and the RLS lie.

Locating areas for sectioning

A wooden artefact is usually defined as a piece of wood that has been modified by human agency, e.g. by a tool or, perhaps, fire. It is essential that any wood removed for identification, however small, does not remove or damage the evidence for the object being an artefact or evidence for the function of the object or nature of its manufacture. Toolmarks are the most common form of evidence, but also burnt areas, paint or other pigment may be present. Do not add to toolmark evidence on an artefact by cutting too deep a facet with a razor blade. If possible mark on a sketch where the samples have been taken to avoid any confusion.

Some areas of a wooden object may be less suitable for identification than others because of decay in the wood structure often caused by damp or dry rotting micro-organisms or fungi. Areas adjacent to metal nails, rivets, hinges, etc. may be encrusted with metal salts and be very difficult to identify.

One of the problems with identifying artefactual wood is that after the direction of the grain and the most likely areas of an object that lie on the TS, TLS and RLS planes have been decided and then areas that contain toolmarks or paint, will face the public in a museum or are rotten and difficult to section have been subtracted, the question of where exactly to take sections still remains. This is often difficult but rarely impossible. The following three guidelines may help:

- Identification can be made on sections that are only a few millimetres across and with practice a hand-held razor blade can take a section that is less than 50–75μm thick. Practice is important since a thick section will not only will leave its own quite visible facet in the wood, it will also be less useful in identification.
- Identification can be made on sections that are less than perfectly parallel to the TS, TLS or RLS planes so where it is impossible to take sections that are perfectly parallel to the transverse and longitudinal, the closest sectionable plane may suffice.
- Where an artefact is likely to be made from wood from a well-known and not too large woody flora (such as that of northern Europe) then identification can often be based on sections from two of the three planes only and very occasionally on the TS alone.

Broken areas on an artefact are ideal for removing thin sections of wood for identification. Since they expose areas of wood that are internal to the original artefact they cannot possibly have toolmarks or any surface layers on them.

Non-artefactual wood

Non-artefactual wood usually lacks all of the complications that beset the identification of artefactual wood. By its nature it has no toolmarks, paint or stains, is unlikely to be put on display in a museum, nor will its interpretation be lost for the sake of a few marks from a razor blade. In most cases a small sample of wood can be taken for identification at a later time. Some of the problems of determining the direction of grain may be faced, however, since usually there is no difficulty in cutting into the wood, grain is often easier to locate.

Sectioning methods

Waterlogged wood and frozen wood

Waterlogged wood is very variable in its preservation, partly as a result of anatomical differences between taxa and partly a result of differing conditions in the waterlogged post-depositional environment. Ideally, for taking a section, it should be firm and relatively hard, but often it can be spongy or crumbly and while it may look as though it would yield good sections, instead these fall into fragments as soon as they slide the razor blade. Some waterlogged wood, especially the heartwood of oak, can be extremely hard and difficult to section.

Waterlogged wood which is a little too soft to section may be hardened by replacing the water in the tissues with alcohol. Simply place a small sample (approximately 1 cm cubed) of the wood to be identified in 30% alcohol or IMS for about four hours. increase the concentration of alcohol to 40%, then 60 % and then 80% each for about two hours. Follow this by 90% alcohol for about two hours and then into 100% for a further two hours. This will normally harden the tissues for sectioning by hand.

This method is obviously not possible with artefactual wood unless the whole object can be placed in alcohol, though often this will take much longer than a few hours to produce the desired result.

Taking a section

Using half of a double-sided razor blade held at one end between your thumb and index finger, gently place the other end on the wood with the blade adjacent to the area to be sectioned. Incline the razor blade very slightly so that when it is drawn across the area from which the section is to be taken it cuts into the wood. Draw the blade across the wood at an angle, lifting towards the end to bring the blade out of the wood (see Figure 1.10). The action is one of pulling the blade across and then out of the wood, rather than one of either sawing or pushing the blade into the wood. The section can be lifted and slid from the razor blade onto a glass microscope slide with a fine paint brush that has been dipped in IMS. A few drops of IMS should be put onto the section and then it should be covered with a cover slip. This should be lowered gently onto the slide from one edge while the other edge rests upon the slide, thus avoiding too many bubbles in the section. This is not a difficult technique, but is one that usually requires practice before perfect sections can be taken.

A great deal of wood preserved by waterlogging has undergone both compression and distortion. Such wood is still possible to identify although many of the most obvious characters, especially when seen in transverse section, are difficult to recognise. Some of the compression and distortion may be lost if the thin section is allowed to sit in IMS for a while after being taken, but this will not always happen. However, all but the most distorted or

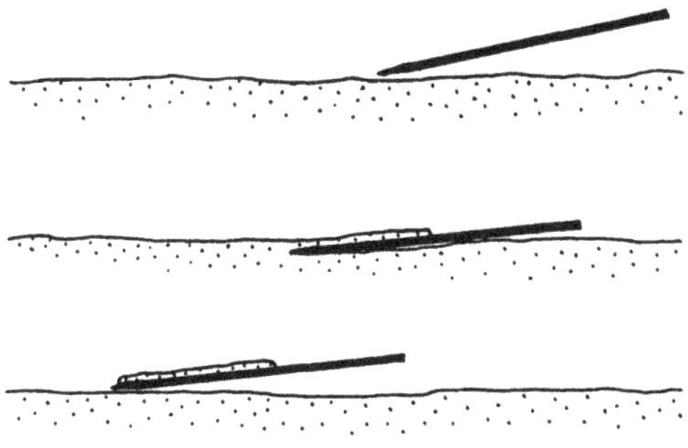

Figure 1.10 Sections through the process of taking a thin section of waterlogged wood using a razor blade (the thickness of the thin section is exaggerated). As the blade enters the wood it must not dig in and go deeper than the desired thickness of the section being taken.

compressed wood should be identifiable, though it will take some time and effort as well as possibly numerous sections in order to work out some character states.

Frozen wood is mostly treated as waterlogged wood once it has defrosted. The condition of the wood once it has defrosted varies depending on the degree of waterlogging and the general condition of the wood prior to freezing, whether the wood has suffered from the growth of ice crystals within its structure and if there has been any episode of thawing and subsequent freezing during the post-depositional period. Methods for identification follow that for waterlogged wood unless the defrosted wood is very dry in which case follow the procedure below.

Desiccated wood

Most desiccated wood will fall into one of three categories:

1. Artefactual or structural wood preserved in continuously dry archaeological contexts such as cave sites, or climatically arid areas of the world. Such wood, though often preserved in marvellous condition, is sometimes a little soft and delicate to the touch. However, such wood can be quite hard and robust and have the feel of wood that is very much younger. As with waterlogged wood this category of wood is very variable.
2. Structural timbers in standing buildings are effectively desiccated. Most would have been used green (ie unseasoned) and then seasoned *in situ*. The preservation of such timbers varies from being hard and robust to soft and crumbly depending on the degree of insect, fungal and micro-organism damage. In northern Europe much structural wood is *Quercus* sp. (oak) within the temperate zone and *Pinus sylvestris* (pine) or *Picea abies* (spruce) in the boreal zone.
3. Waterlogged archaeological artefacts that have been conserved and non-archaeological wooden artefacts from ethnographic or historic collections are also treated in the same way as desiccated wood .

As with waterlogged wood desiccated wood is very variable. Its state of preservation is dependent on species but also on the conditions of preservation. Before a section is taken the location to be sectioned should have a few drops of IMS placed upon it so that the sliver of wood, when lifted, is wet. This is particularly important for desiccated wood that is either soft or friable – sections will hold together more easily if made wet with IMS.

Taking a section

A section may taken as an elongated and very shallow saucer-shaped sliver of wood using a shallowly flexed razor blade. After it has dug into the wood it begins to lift out the other side. The resulting section may be slightly thick in the middle but will be much thinner around the edges. The section will leave a very shallow depression of much less than 1mm.

On a perfectly flat longitudinal surface (to section RLS or TLS) the edge of the razor blade can be pushed at a very low angle either along the grain or across it. If sectioning along the grain it is possible to dig into the wood if the blade is held at too high an angle. Sectioning across the grain could result in the section breaking into longitudinal strips if the wood is friable, but if sectioned well the section should be uniformly thick and leave practically no mark.

Charcoal

Charcoal is unlike other forms of archaeologically preserved wood. The act of preservation by carbonisation is a chemical transformation that results in a material that is quite inert but at the same time often very fragile. Because of its natural fragmentation properties, the physical rigours of the post-depositional environment, and often because of rather drastic recovery techniques, fragments of charcoal greater than 1 to 1½ cm across are rarely preserved. Most fragments of charcoal are a great deal smaller.

The identification of charcoal is a very different technique compared to the identification of wood preserved by the other methods described above. Although it is possible to take a thin section by embedding the charcoal fragment in resin and grinding it down until it is thin enough for light to pass through, this is a long and laborious method. The use of pressure fracturing to create the three planes, TS, TLS and RLS is very much quicker and is relatively less destructive since though the original fragment is broken the two resulting halves of a fragment resulting from a pressure fracture remain for further fracturing.

As with wood preserved by other methods, the first task is to orient the fragment. This is best done under low power reflected light microscopy. Charcoal is a highly reflective material and so with clean charcoal this is often easy since the light reflected from the linearly oriented cells of the RLS and TLS is very different from the way it is reflected from the TS. This task is less easy if the charcoal surface is abraded, covered with dust or mineral concretions. Carefully blowing with pressurised air will remove some of the dust, but even this will not always allow you to orient the fragment.

If the direction of grain is obvious then a pressure fracture may be made to provide a clean surface, since what appear to be clean surfaces on unfractured charcoal will be slightly dusty and abraded, obscuring diagnostic characters. The first pressure fracture should be across the grain revealing a fresh transverse surface. A pressure fracture is made by placing a sharp and rigid blade onto the charcoal surface to be fractured (see Figure 1.11).

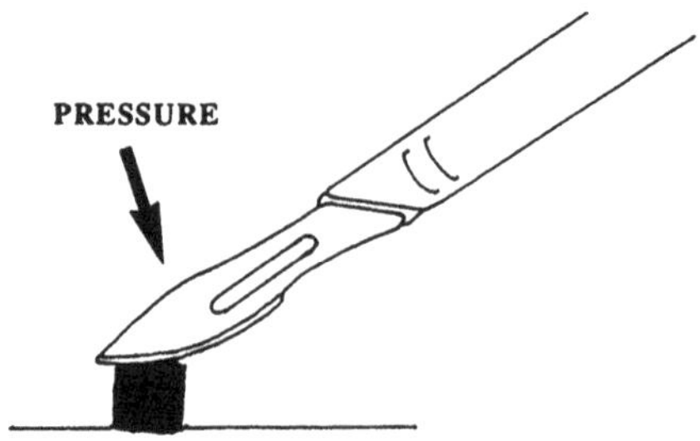

Figure 1.11 Pressure fracturing of a charcoal fragment using a scalpel blade.

Applying slight pressure to the top of the blade will cause the charcoal to split into two pieces. It is important not to slice, cut or saw your way through the charcoal – simply apply enough pressure for the charcoal to split into two without the blade cutting through the charcoal at all. From this transverse surface pressure fractures can be made in the direction of the grain that reveal the radial longitudinal surface and the tangential longitudinal surface. If the direction of the grain is not apparent then a pressure fracture must be made so that a clean surface can be observed. This should reveal sufficient indication of the grain to suggest the locations of the sections required.

Fragments with their clean fresh fractured surfaces uppermost should be mounted in sand and viewed under high power reflected light microscopy (epi-illuminated microscopy).

Reference collections

Although this book aims to provide sufficient information for wood to be identified, it is strongly recommended that modern reference material is also examined for two reasons. The first is to be able to take account of the inherent variability of wood (described above) which is only possible through the examination of a large number of samples from a range of growing sites and positions within a woody plant. The second reason, important for those beginning identification, is that the only way to familiarise yourself with the three-dimensional organisation of wood anatomy is by spending many hours looking at thin sections through a microscope. It is difficult to gain this familiarity by looking at photographs in this or any other book.

Concluding remarks

This guide is a summary of methods only and it is important to remember that practice alone will improve your skills in selecting the correct location for sectioning and in taking the section itself. Modern softwood is harder than most archaeologically preserved woods but is suitable for practice.

When identifying artefacts it is important that the curator of any collection is informed of the sampling procedure prior to any sections being taken. An uninformed curator finding a wood analyst with an artefact in one hand and a razor blade in the other is unlikely to be sympathetic. It is often wise to demonstrate the procedure to a curator on a piece of modern softwood – the small section taken usually reassures.

How to use this book to identify wood

This volume is part atlas and part diagnostic key. Its intention is to teach the identification of the northern European woody flora rather than simply to provide an identification by following individual stages of a key. In this respect there are small keys to groups of taxa, starting with the key to anatomical groups at the beginning of Part 2. Further than this, readers will have to read the descriptions of individual taxa to decide on an identification. Tables comparing characters of taxa with similar anatomies are given as well as suggestions of taxa that may provide possible confusion. Above all, taxa that are most similar to each other are placed adjacent to each other so that direct comparisons can be made.

To begin take a transverse section (or fracture plane for charcoal) of the wood you need to identify, turn to the key to anatomical groups (pp. 26–7) and place it within one of the groups. Go to the group page and follow the short key to taxa groups and follow from there.

If identification is problematical the most important diagnostic characters are shown in tables in Part 3. Here, all taxa that share range of character are listed.

Bibliography

Fitter, A. (1978) *An Atlas of the Wild Flowers of Britain and Northern Europe*. London: Collins

Jalas, J., J. Suominen and R. Lampinen (eds) (1973–1996) *Atlas Florae Europaeae: distribution of vascular plants in Europe*. Helsinki: Committee for Mapping the Flora of Europe and Societas Biologica Fennica, Vanamo.

Jane, F. W. (1956) *The Structure of Wood*. London: A and C Black.

Phillips, E. W. J. (1948) Identification of softwoods by their microscopic structure. *Forest Products Research Bulletin* 22. London: HMSO.

Schweingruber, F. H. (1990) *Anatomy of European Woods*. Stuttgart: WSL and Haupt.

Wheeler, E. A., P. Bass and P. E. Gasson (1989) IAWA list of microscopic features for hardwood identification. *IAWA Bulletin* **10**, 219–332.

Wilson, K. and D. White 1986 *The Anatomy of Wood: its diversity and variability*. London: Stobart and Son.

PART TWO

Identification of the Northern European Woods

Index to taxa by group

Alphabetical index to taxa

Index to maps

Organisation of the taxa descriptions and key to the anatomical groups

Rather than organise the taxa described in this book by taxonomic or alphabetical order, they are placed in groups that share common anatomical features. It is hoped that such an organisation will help the user to more easily identify an unknown wood by being able to place it within a small group of taxa that are anatomically similar. Diagnostic characters that differentiate members of each group can then be determined and the wood identified without anatomically dissimilar woods that happen to be taxonomically related or beginning with the same letter of the alphabet cluttering up the process. It should also aid in the identification of archaeological material where not all characters are visible.

The anatomical characters chosen to place the taxa into groups are, at the highest level, the arrangement of vessels or pores seen in transverse section. Below this, ray width, a character seen in TLS, is used. The Gymnosperms are dealt with separately as they form a small and anatomically discrete group alone.

Set out below is a simple key that should be used to identify the anatomical group to which a taxon belongs. The characters are all identifiable from the transverse section alone although other sections will have to be taken to identify the taxon further.

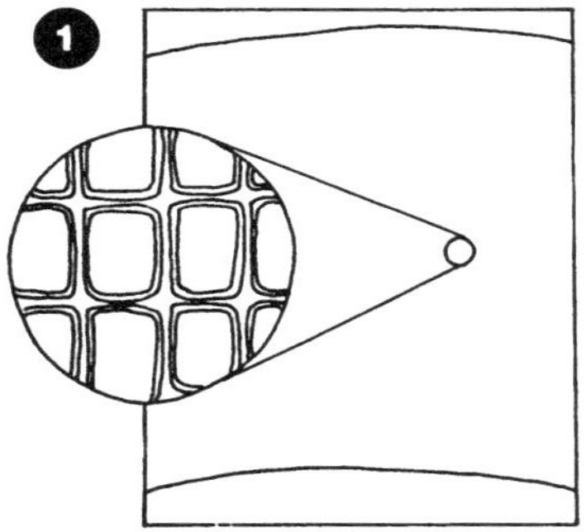

1 Are there vessels present in the transverse section? Vessels are visible as round holes, or pores, in the transverse section and though they vary greatly in size they are always larger than the cells that surround them. Gymnosperms lack vessels – their tissue is made up of many small, evenly sized tracheids though the walls of these may get thicker during the growth of one year (see Figure [1]).

- vessels absent – **Group 1, Gymnosperms** → page 30
- vessels present (Dicotyledons) → 2

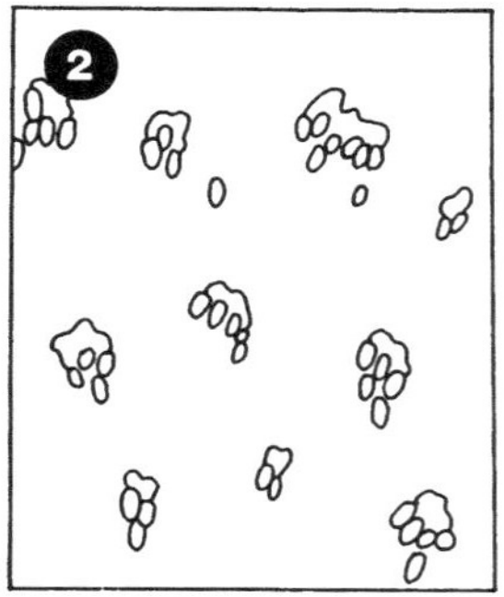

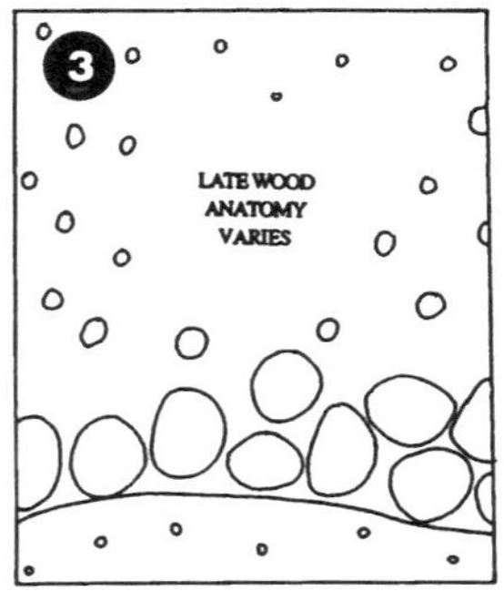

2 Wood with vessels
Tissue made up of irregularly concentric rings of bundles of vessels in small radial files and clusters each placed internally to bundles of phloem. Rays virtually absent (one species described) (see Figure [2]).

- **Group 2, Wood with included phloem** → page 44
- Tissue made up of annual rings of tissues containing vessels within a ground tissue but without internal phloem → 3

3 Is there a row of vessels at the beginning of a year's growth that is much larger than the vessels that are present in the rest of the year's growth (see Figure [3])? The organisation of the vessels after the initial ring of early wood vessels varies considerably.

- Vessels present in the early wood much larger than in the later wood – **Group 3, Ring porous woods** → page 46

– Are the vessels much the same size throughout the year's growth (they may get slightly smaller towards the end of the year) though the vessels may be arranged in clusters, radial files, or are solitary?

- Vessels much the same size throughout the year (diffuse porous woods) → **4**

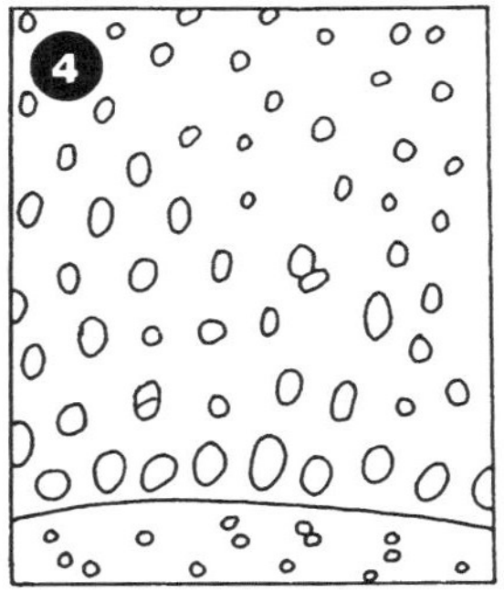

4a Diffuse porous woods. The vessels all more or less solitary. There may be the occasional radial, tangential or oblique groups of two or at the most three vessels (see Figure [4]).

- **Group 4, Diffuse porous with solitary vessels** → page 72

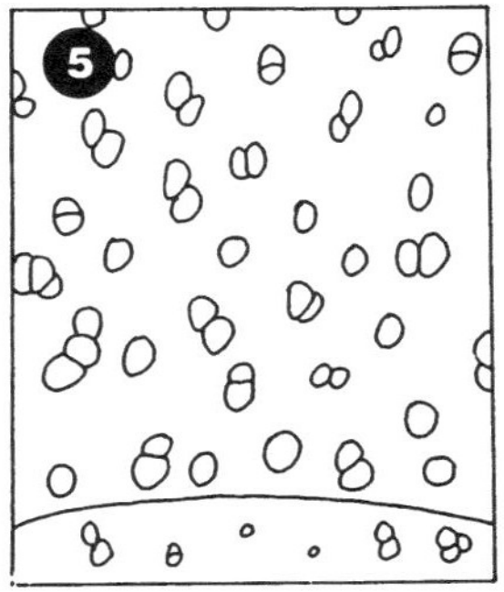

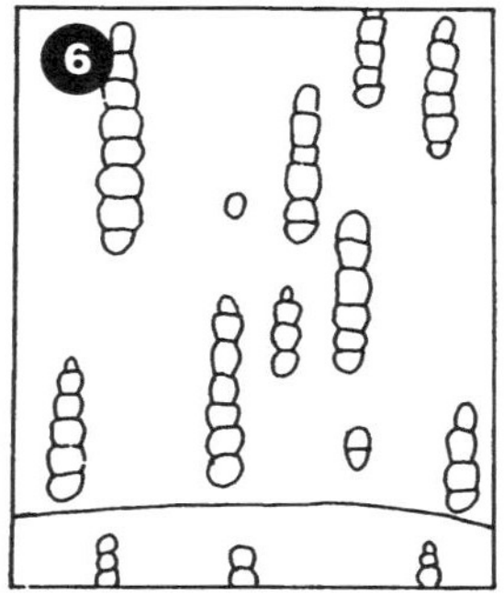

4b Diffuse porous woods. There are solitary vessels but at least half of vessels are in clusters, short radial files or tangential/oblique bands of two to three vessels (see Figure [5]).

- **Group 5, Diffuse porous with clusters and/or short radial files** → page 102

4c Diffuse porous woods. There are solitary vessels but most are in long radial files of three to five vessels or more.

- Most of the vessels are part of radial files (see Figure [6]) – **Group 6, Diffuse porous with long radial files** – page 138

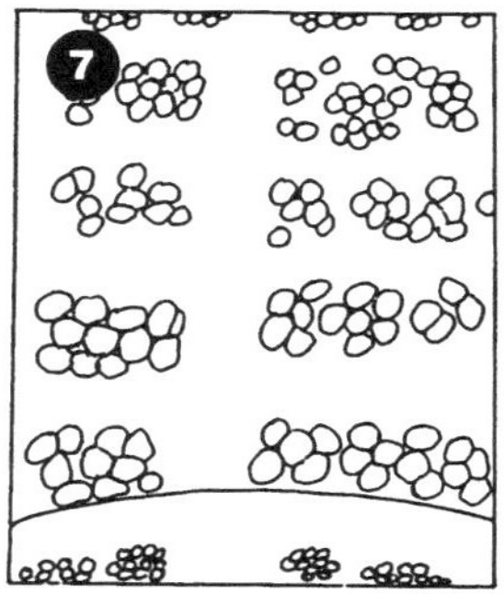

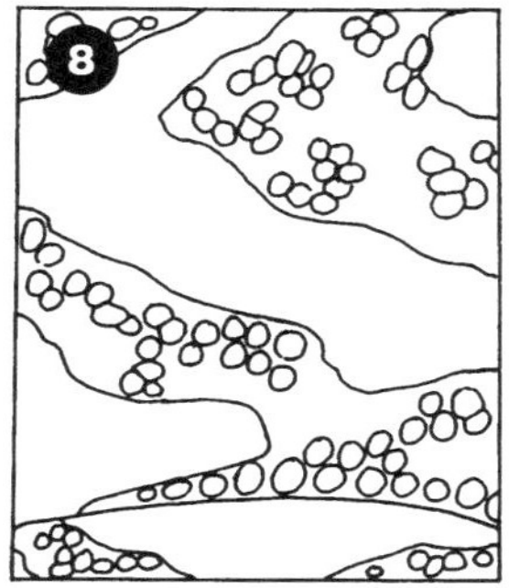

4d Diffuse porous woods. Vessels are mostly grouped in clusters of ten or more vessels. These are often oriented tangentially. Vessels within clusters may appear to be solitary (see Figure [7]).

- **Group 7, Diffuse porous with clusters** – page 149

4e Diffuse porous woods. Vessels are grouped into a dendritic arrangement within which vessels may appear to be solitary or in clusters (see Figure [8]).

- **Group 8, Diffuse porous with pores in dendritic pattern** – page 156

A note on the tables

The tables within this part and Part 3 use a number of rather idiosyncratic abbreviations, as follows.

- **Ray size, width and height** This is expressed in cell number, i.e. 1 is unisereate, 2 bisereate and so on. Larger cell number width and heights are approximate such that if a figure of 50 is given for height, a ray of 55 cells high might well be found though these will not be common.
  - The width of rays for the Gymnosperms is always given as unisereate since these are by far the most common though bisereate rays are not unusual in most taxa.
  - Large rays are expressed in widths and heights of millimetres or centimetres.

- **Ray anatomy** Where space has not allowed the terms in full, 'Ht' has been used for heterogeneous and 'Hm' for homogeneous.

- **Presence and absence**
  - **Present** is given where a character state is present, **absent** when the character state is absent **(present)** is given where a character state is rare, **very rare** is given where a character state is unusual but has been recorded. If a particular character is listed in a table (rather than its presence/absence) then if it is rare it will appear within brackets. This is often the case with ray widths and heights where rare large or small rays may be present occasionally.

In tables that deal with the ring porous woods, within their own section and within other sections where their late wood anatomy is compared with the anatomy of diffuse porous woods, the following abbreviations that relate to the arrangement of vessels are used:

S solitary
SRF short radial files
C clusters
TC tangential clusters

A note on the distribution maps

The maps presented in Part 2, alongside each of the anatomical descriptions, have been compiled with reference to *Atlas Florae Europaeae* (Jalas *et al.* 1973–1996) and Fitter (1978). For each map the natural distribution is marked as a solid black area. Where the distribution of an introduced taxon is given the text will indicate whether this is a modern or archaeological distribution. In some cases knowledge of the full native distribution is patchy. In such cases where there is a presumed distribution this is marked as a stippled area. These maps are designed to give a broad indication of distribution and are not detailed enough to indicate local variations in distribution. For such detail local floras should be consulted.

Group 1 – Gymnosperms

In both the alpine region of central Europe and the boreal region of northern Europe the Gymnosperms form dominant woodlands. Both ecologically and economically the Gymnosperms have been and still are important taxa.

Table 2.1 Morphology of the gymnosperm taxa

Abies alba	large tree
Larix decidua	large tree
Picea abies	large tree
Pinus sylvestris	large tree
Pinus mugo	small tree
Juniperus communis	small tree/shrub
Juniperus sabina	small tree
Taxus baccata	Medium tree

General anatomy

The Gymnosperms have a generally simple anatomy with relatively few diagnostic characters. The degree of abruptness of the change from early to late wood varies but is not always a reliable diagnostic character. Presence of single characters can be of high diagnostic value, for example: spiral thickening in *Taxus*, parenchyma cells with coloured contents in *Juniperus* and dentate walls in the ray tracheids in *Pinus*. The height of rays and the presence of resin ducts is of high diagnostic value as are the ray pits, though these may be difficult to see in archaeological material.

Table 2.2 Summary of the anatomical characters of the Gymnosperms

Taxon	Ray height	Resin ducts present	Spiral thickening present	Ray tracheids present	Early wood ray pits	Late wood ray pits
Abies alba	15–25(–40)	no	no	no	taxodioid	piceoid
Larix decidua	10–15(–25)	yes	no	yes	piceoid	piceoid
Picea abies	10–15(–25)	yes	no	yes	piceoid	piceoid / (cupressoid)
Pinus sylvestris	8–15	yes	no	yes	fenestriform	fenestriform
Juniperus communis	2–5(–15)	no	no	no	small cupressoid	cupressoid
Taxus baccata	5–12(–20)	no	yes	no	cupressoid	cupressoid

Taxa descriptions

Group key

Spiral thickening prominent (LS)	*Taxus baccata*
Ray height 15 or less cells high (TLS)	
• Resin ducts present	*Pinus sylvestris/P. mugo*
• Resin ducts absent (TLS)	*Juniperus communis/J. sabina*
Ray height often over 15 cells high (TLS)	
• Resin ducts absent (TS/TLS)	*Abies alba*
• Resin ducts present (TS/TLS)	*Picea abies, Larix decidua*

Spiral thickening prominent (LS)

Taxus baccata – Yew (Cupressaceae)

Plate 1

Taxus is a low, though often broadly spreading tree with a central northern European, southern Scandinavian and western European distribution (see Map 1).

Taxus baccata has a relatively simple anatomy, with no resin ducts, no parenchyma and homocellular rays. The transition from early to late wood is gradual. Rays are up to 20 cells high, though commonly less. Ray pits are cupressoid. The most diagnostic character is the distinctive spiral thickening in tracheids seen in LS.

This is the only Gymnosperm taxon present in northern with Europe distinct spiral thickening. Only *Abies alba* and *Juniperus* spp. share the character of having no resin ducts.

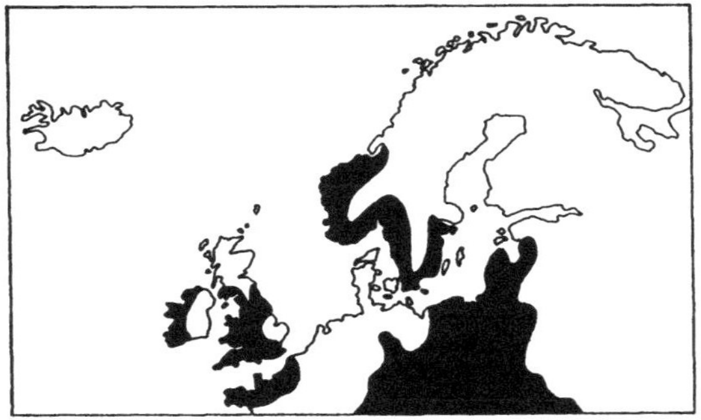

Map 1 Distribution of *Taxus baccata*

Table 2.3 Anatomical characters of *Taxus baccata* (Cupressaceae)

Character	*Taxus baccata*
TS	
Early/late wood transition	gradual
Axial resin canals	absent
Epithelial cells	–
Parenchyma	absent
TLS	
Ray height	5–12 (–20)
Radial resin canals	absent
Spiral thickening	present
RLS	
Axial tracheid pits	uniseriate
Crassulae	absent
Ray tracheids	absent
Ray tracheid wall	–
Nodular tangential ray walls	absent
Early wood ray pits	cupressoid
Late wood ray pits	cupressoid
Crystals in ray cells	absent

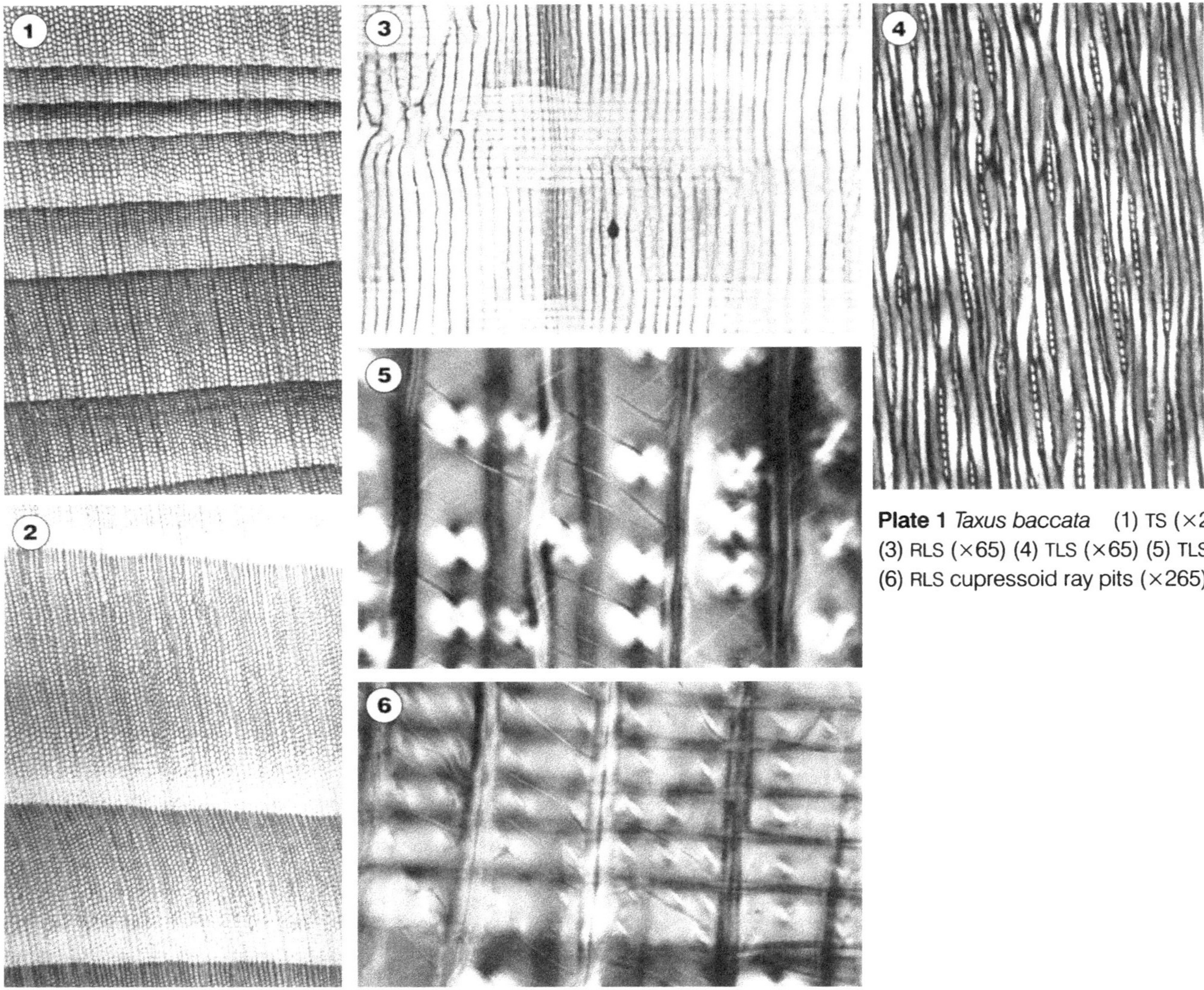

Plate 1 *Taxus baccata* (1) TS (×28) (2) TS (×28) (3) RLS (×65) (4) TLS (×65) (5) TLS spiral thickening (×265) (6) RLS cupressoid ray pits (×265).

Ray height 15 or less cells high (TLS), Resin ducts present in rays (TLS)

Pinus sylvestris – Pine (Scots pine) (Pinaceae)

Pinus mugo – Mountain pine (Pinaceae)

Plate 2

Pinus sylvestris is a tall forest tree of the boreal region with a distribution that extends down into northern Europe (see Map 2). *Pinus mugo* is a somewhat smaller tree with a more central European distribution extending into northern Europe (see Map 3). There are many more pines in Europe that can be differentiated on the basis of their anatomy although these two species cannot.

The transition between the early and late wood in *Pinus sylvestris* and *Pinus mugo* is rather abrupt. Resin ducts with thin-walled epithelial cells occur in the late wood and in rays. Rays are 8–15 cells high and heterocellular. Ray tracheids have distinctive internal dentate walls. Ray pits are fenestriform.

The presence of both dentate walls in the ray tracheids and the fenestriform ray pits should prevent confusion with any other woods.

Map 2 Distribution of *Pinus sylvestris*

Table 2.4 Anatomical characters of *Pinus sylvestris* and *P. mugo* (Pinaceae)

Character	*Pinus sylvestris/ P. mugo*	Character	*Pinus sylvestris/ P. mugo*
TS		RLS	
Early/late wood transition	abrupt	Axial tracheid pits	uniseriate
Axial resin canals	throughout	Crassulae	absent
Epithelial cells	thin-walled	Ray tracheids	present
Parenchyma	absent	Ray tracheid wall	dentate
TLS		Nodular tangential ray walls	absent
Ray height	8–15	Early wood ray pits	fenestriform
Radial resin canals	present	Late wood ray pits	fenestriform
Spiral thickening	absent	Crystals in ray cells	absent

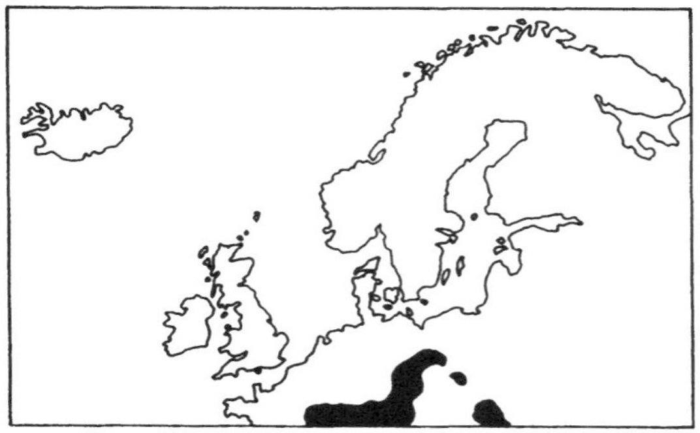

Map 3 Distribution of *Pinus mugo*

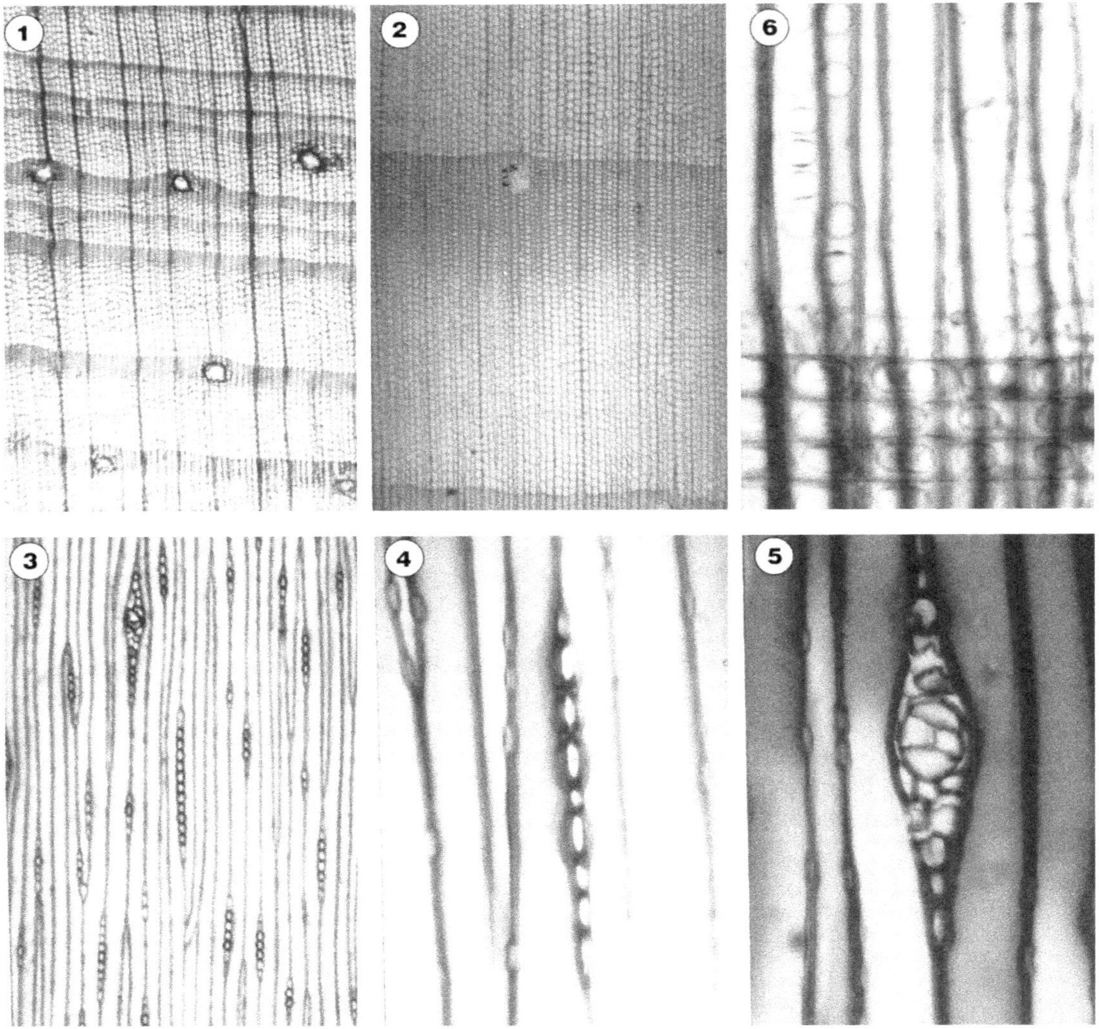

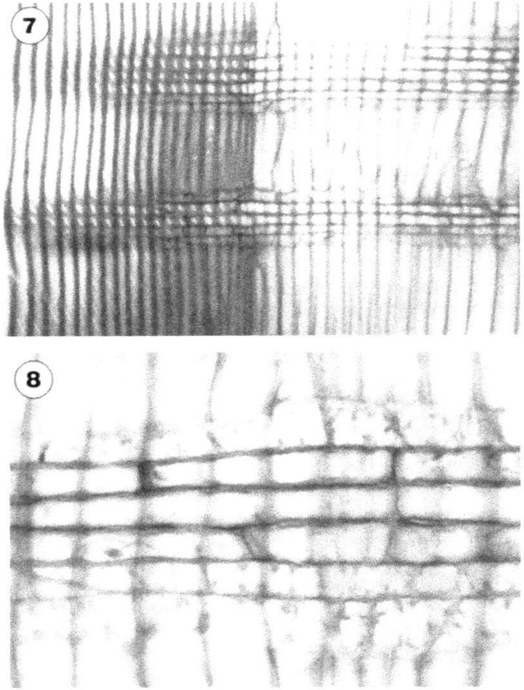

Plate 2 *Pinus sylvestris* (1) TS (×28) (2) TS (×28) (3) TLS (×65) (4) TLS (×140) (5) TLS ray resin duct with thin-walled epithelial cells (×140) (6) RLS ray tracheid with dentate walls (×140) (7) RLS (×65) (8) RLS ray with fenestriform ray pits and ray tracheid with dentate walls (×140).

Ray height 15 or less cells high (TLS), Resin ducts absent (TLS)

Juniperus – Junipers (Cupressaceae)

Plate 3

There are many species of *Juniper* in Europe, many of them in the Mediterranean. Two species have distributions in northern Europe, *J. communis*, the common juniper which grows throughout the region (see Map 4) and *J. sabina* which just extends into the region from a broader distribution further south (see Map 5). None of the *Juniper* species can be differentiated on the basis of their wood anatomy.

The anatomy of *Juniperus* is relatively simple. There are no resin ducts, and rays are homocellular. Parenchyma cells with coloured contents are present and easily seen in the transverse section. Rays are generally 2 to 5 cells high but occasionally higher. Ray pits are cuppressoid and decrease in size over a year's growth.

The lack of resin ducts, ray height and presence of parenchyma cells should prevent confusion with other woods, however see *Abies alba*.

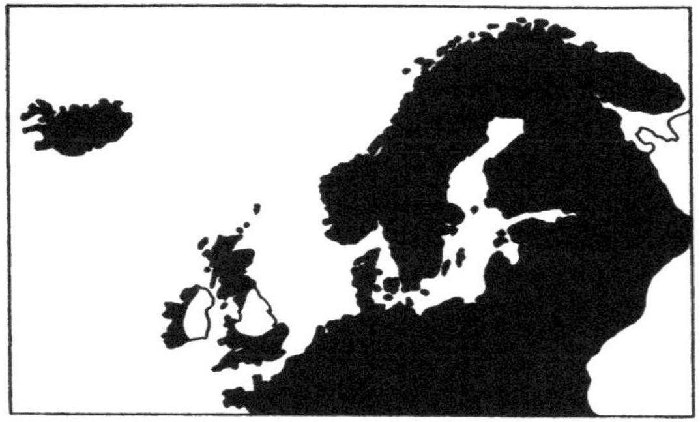

Map 4 Distribution of *Juniperus communis*

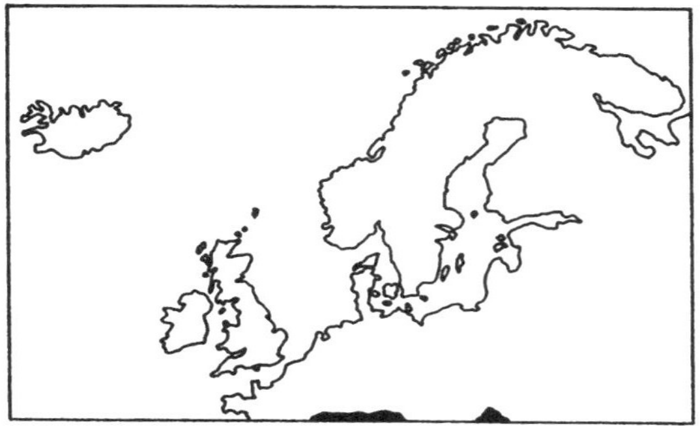

Map 5 Distribution of *Juniperus sabina*

Table 2.5 Anatomical characters of *Juniperus* (Cupressaceae)

Character	*Juniperus communis/ J. sabina*	Character	*Juniperus communis/ J. sabina*
TS		RLS	
Early/late wood transition	gradual	Axial tracheid pits	uniseriate
Axial resin canals	absent	Crassulae	absent
Epithelial cells	–	Ray tracheids	absent
Parenchyma	present	Ray tracheid wall	–
TLS		Nodular tangential ray walls	present
Ray height	2–5(–18)	Early wood ray pits	cupressoid
Radial resin canals	absent	Late wood ray pits	small cupressoid
Spiral thickening	absent	Crystals in ray cells	absent

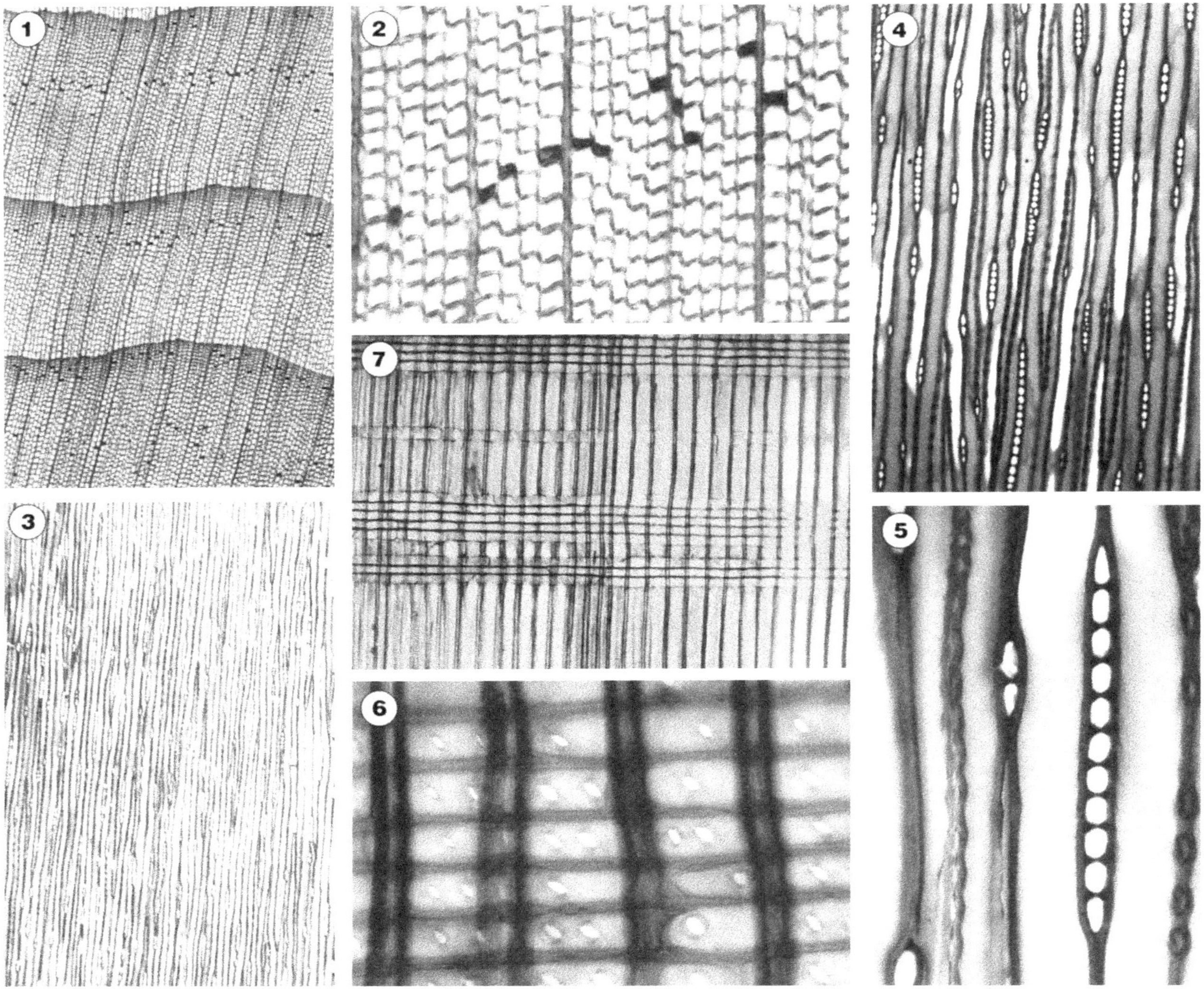

Plate 3 *Juniperus communis*
(1) TS (×28) (2) TS parenchyma cells with coloured contents (×65) (3) TLS (×28)
(4) TLS (×65) (5) TLS (×265)
(6) RLS (×265)
(7) RLS cupressoid ray pits (×65).

Ray height often over 15 cells high (TLS), Resin ducts absent (TS/TLS)

Abies alba – Silver fir (Pinaceae)

Plate 4

Abies alba is a tall forest tree of central Europe with a distribution that extends into northern Europe (see Map 6).

The anatomy of *Abies alba* is relatively simple. It has no resin ducts, and rays are homocellular. The transition from early to late wood may be abrupt or gradual. Rays are generally between 15 and 25 cells high though often higher. Ray cells often contain crystals and have nodular tangential walls. Ray pits are taxodioid in the early wood and piceoid in the late wood.

Abies alba could be confused with *Juniperus* though the higher rays and nodular tangential ray walls are diagnostic.

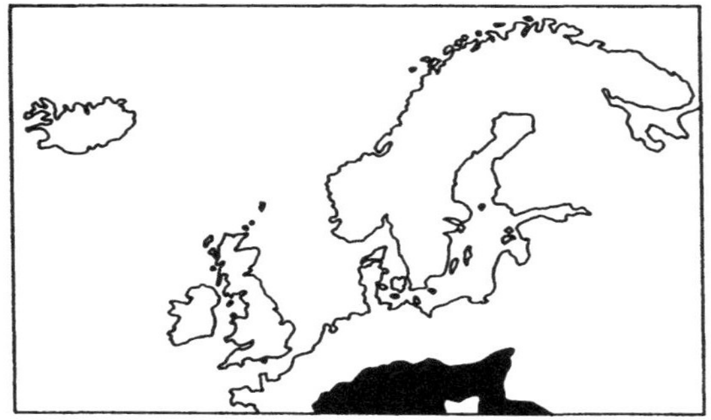

Map 6 Distribution of *Abies alba*

Table 2.6 Anatomical characters of *Abies alba* (Pinaceae)

Character	*Abies alba*
TS	
Early/late wood transition	gradual/abrupt
Axial resin canals	± absent
Epithelial cells	–
Parenchyma	(rare terminal)
TLS	
Ray height	15–25(–40)
Radial resin canals	absent
Spiral thickening	absent
RLS	
Axial tracheid pits	uniseriate (bisereate)
Crassulae	absent
Ray tracheids	absent
Ray tracheid wall	–
Nodular tangential ray walls	present
Early wood ray pits	taxodioid
Late wood ray pits	piceoid
Crystals in ray cells	(present)

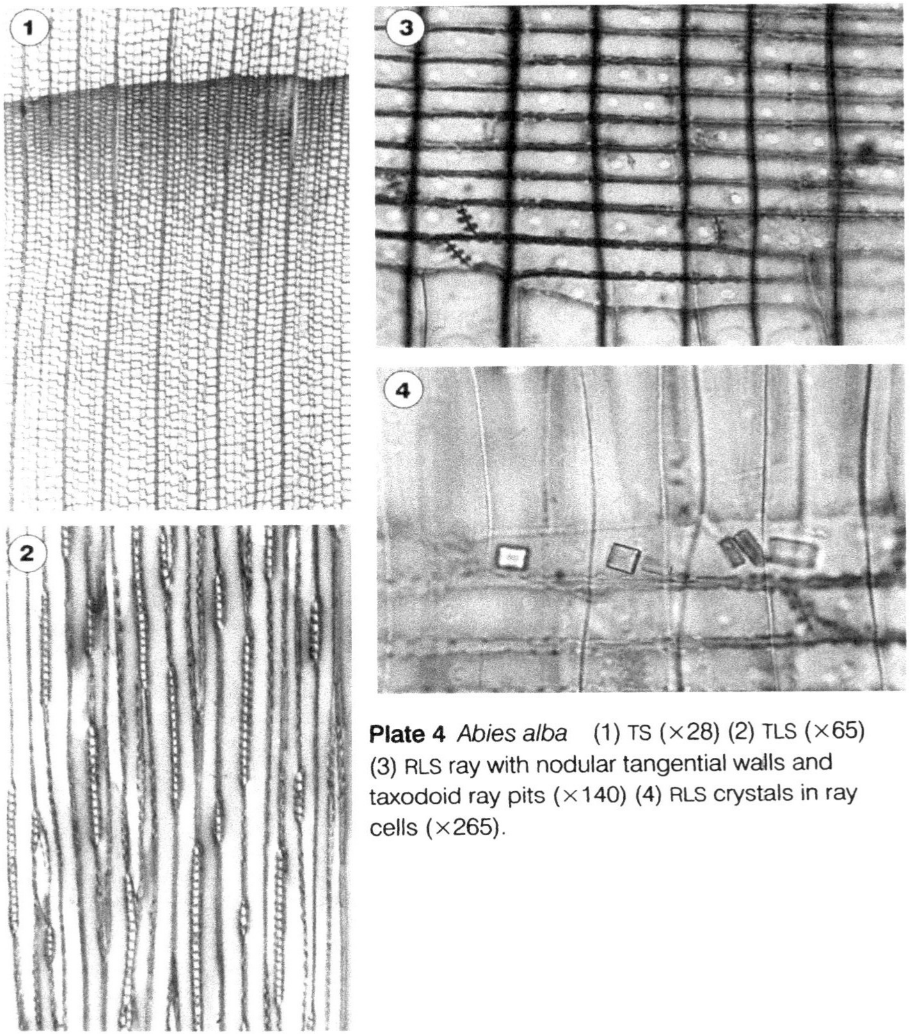

Plate 4 *Abies alba* (1) TS (×28) (2) TLS (×65) (3) RLS ray with nodular tangential walls and taxodoid ray pits (×140) (4) RLS crystals in ray cells (×265).

Ray height often over 15 cells high (TLS), Resin ducts present (TS/TLS)

Larix decidua – Larch (Pinaceae)

Picea abies – Spruce (Pinaceae)

Plate 5 and Plate 6

Larix decidua has a central European distribution that extends into Northern Europe (see Map 7). *Picea abies* has a northerly boreal distribution that extends down into northern and central Europe (see Map 8). The two species can be differentiated though this is not always possible. For this reason they are being described together.

The transition between early wood and late wood in *Larix* is abrupt while it is gradual in *Picea*. This character is so affected by environmental factors that it is not of high diagnostic value. In both taxa there are resin ducts with thick-walled epithelial cells in the late wood and in the rays. Rays are heterocellular with smooth-walled ray tracheids present. Rays are up to 15 cells high in both, but may occasionally be higher. Ray pits are piceoid, though occasionally cupressoid in the early wood if *Picea*.

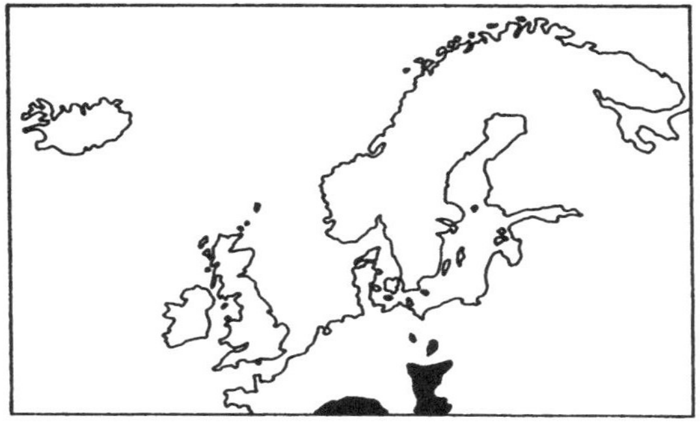

Map 7 Distribution of *Larix decidua*

Differentiation between Larix *and* Picea

In *Larix* bordered pits seen in RLS are frequently in pairs and have between them thickened bars, known as crassulae or 'Bars of Sanio' (see Figure 1.3). These are absent in *Picea*. Schweingruber (1990) also notes that the resin ducts in rays are eccentrically positioned in *Larix* and centred in *Picea*. In poor specimens it is often not possible to differentiate *Larix* and *Picea* though their natural distributions do not overlap.

In poor specimens it may be possible to confuse *Picea* and *Larix* with *Pinus sylvestris/ P. mugo*, however, resin ducts in *Pinus* have thin-walled epithelial cells and Pinus also has dentate ray tracheids and fenestriform pits.

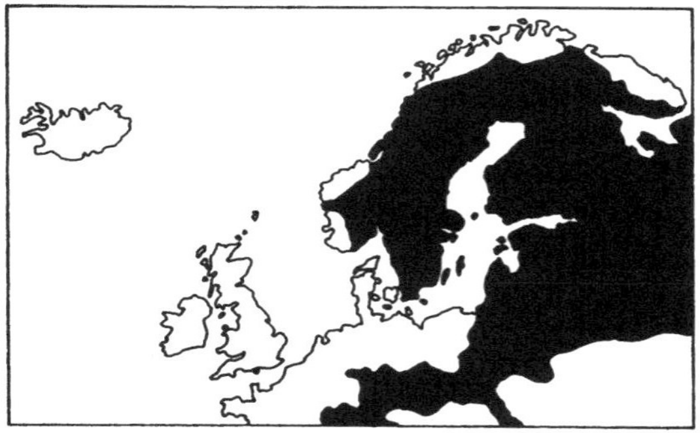

Map 8 Distribution of *Picea abies*

Table 2.7 Anatomical characters of *Larix decidua* and *Picea abies* (Pinaceae)

Character	*Larix decidua*	*Picea abies*
TS		
Early/late wood transition	abrupt	gradual
Axial resin canals	in late wood	in late wood
Epithelial cells	thick-walled	thick-walled
Parenchyma	absent	absent
TLS		
Ray height	10–15(–25)	10–15(–25)
Radial resin canals	present	present
Spiral thickening	absent	absent
RLS		
Axial tracheid pits	bisereate (uniseriate)	uniseriate (bisereate)
Crassulae	present	absent
Ray tracheids	present	present
Ray tracheid wall	smooth	smooth
Nodular tangential ray walls	present	present
Early wood ray pits	piceoid	piceoid (cupressoid)
Late wood ray pits	piceoid	piceoid
Crystals in ray cells	present	present

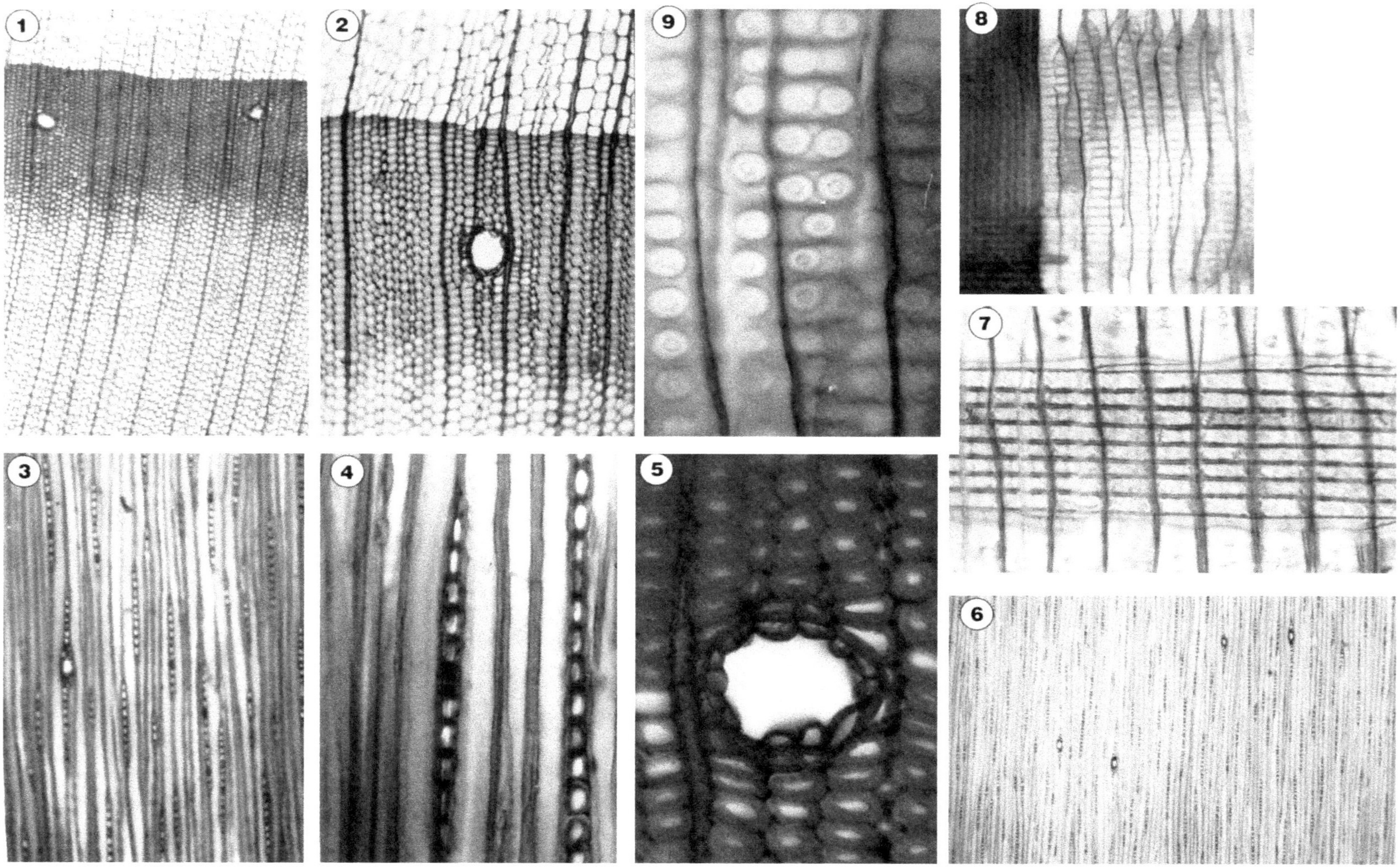

Plate 5 *Larix decidua* (1) TS (×28) (2) TS (×65) (3) TLS (×65) (4) TLS (×140) (5) TS axial resin duct with thick-walled epithelial cells (×265) (6) TLS (×28) (7) RLS ray with piceoid ray pits and smooth-walled pitting (×140) (8) RLS bordered pit pairs and crassulae (×65) (9) RLS pit pairs and crassulae (×265).

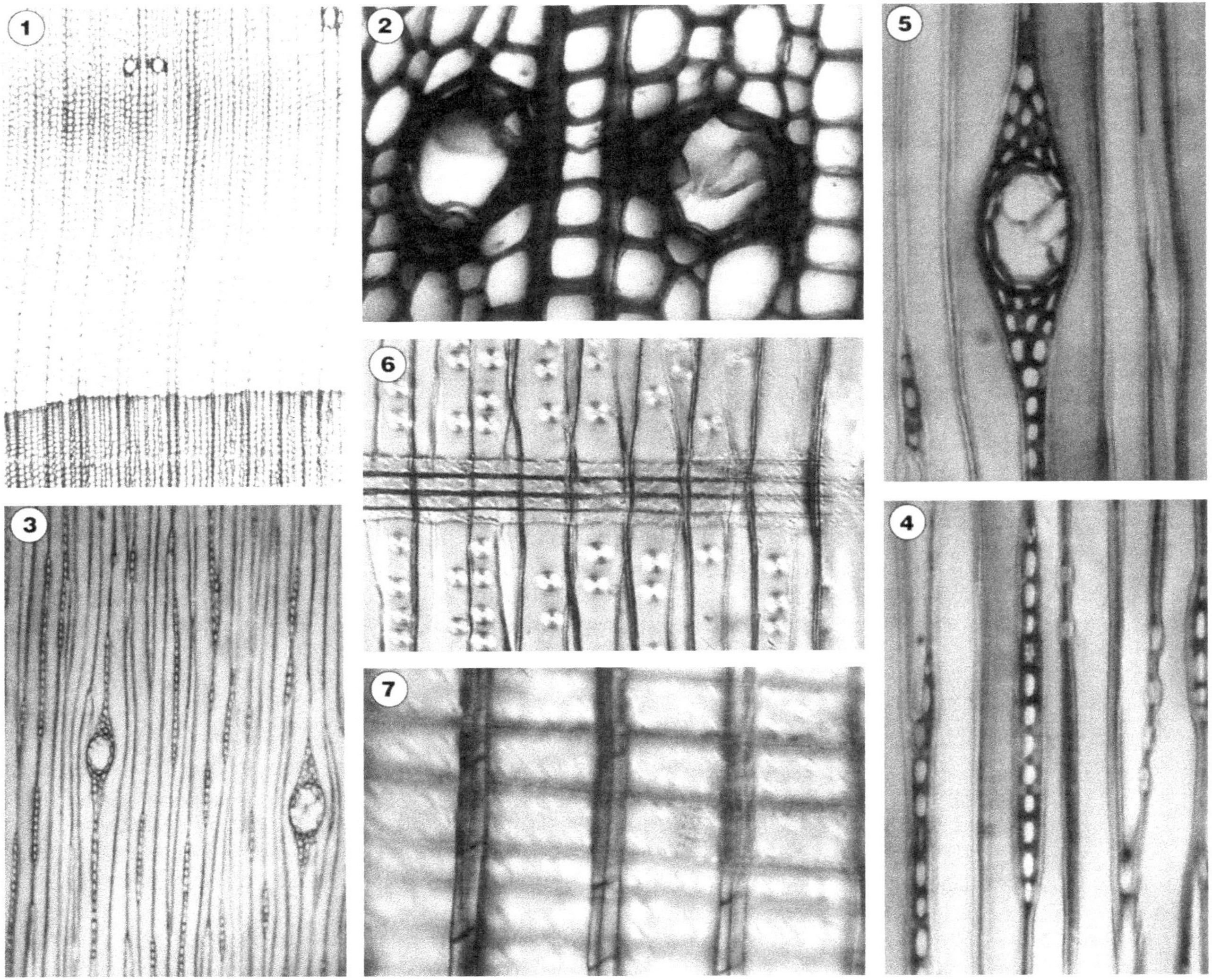

Plate 6 *Picea abies*
(1) TS (×28) (2) TS resin ducts with thick-walled epithelial cells (×265) (3) TLS (×65) (4) TLS (×140) (5) RLS ray resin duct with thick-walled epithelial cells (×140) (6) RLS ray and unisereate bordered pits (×140) (7) RLS ray with piceoid ray pits (×265).

Group 2 – Woods with included phloem

There is a single taxon described here that has included phloem. This is a character of the Chenopodiaceae, a family that is not well represented by wood plants in northern Europe. *Atriplex portulacoides* becomes slightly woody but not sufficiently to be of any economic use. *Suaeda vera* is a medium to large evergreen shrub with a more or less coastal distribution in southern Britain and northern France (see Map 9).

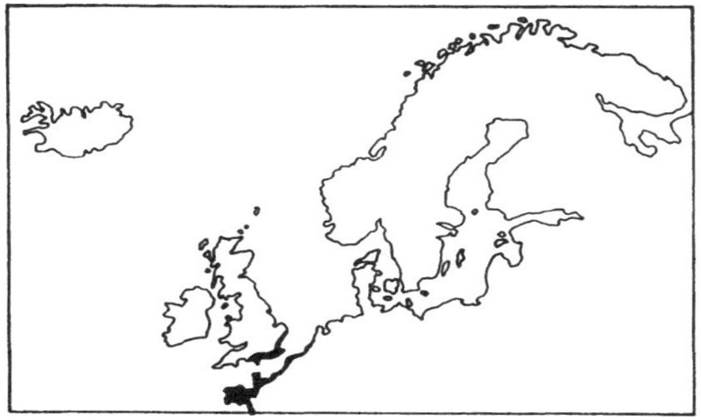

Map 9 Distribution of *Suaeda vera*

General anatomy

Included phloem is a character that is common throughout the Chenopodiaceae and refers to areas of phloem within the secondary xylem. Wood develops in the conventional way resulting in concentric growth rings but these are often indistinct since the vessel/phloem groups are undulating and often discontinuous.

Table 2.8 Anatomical characters of *Suaeda vera* (Chenopodiaceae)

Character	*Suaeda vera*
TS	
Vessel arrangement/ grouping	with included phloem: concentric type
Parenchyma	apotracheal
Ground tissue	–
Ring boundary	undulating
Rays	absent/rare – indistinct
TLS	
Ray width	–
Ray height	–
Sheath cells	–
RLS	
Ray anatomy	–
Libriform fibres	present
Fibre tracheids	absent
Vascular tracheids	present
Spiral thickening	present
Crystals	present
Perforation plates	simple

Suaeda vera – Shrubby sea-blite (Chenopodiaceae)

Plate 7

The xylem /phloem groups are of a concentric type and form a nearly continuous ring. Vascular cells, parenchyma and tracheids are storied. Rays are indistinct or absent. Perforation plates are simple and crystals are often present in the parenchyma. Spiral thickening is present in narrower vessels.

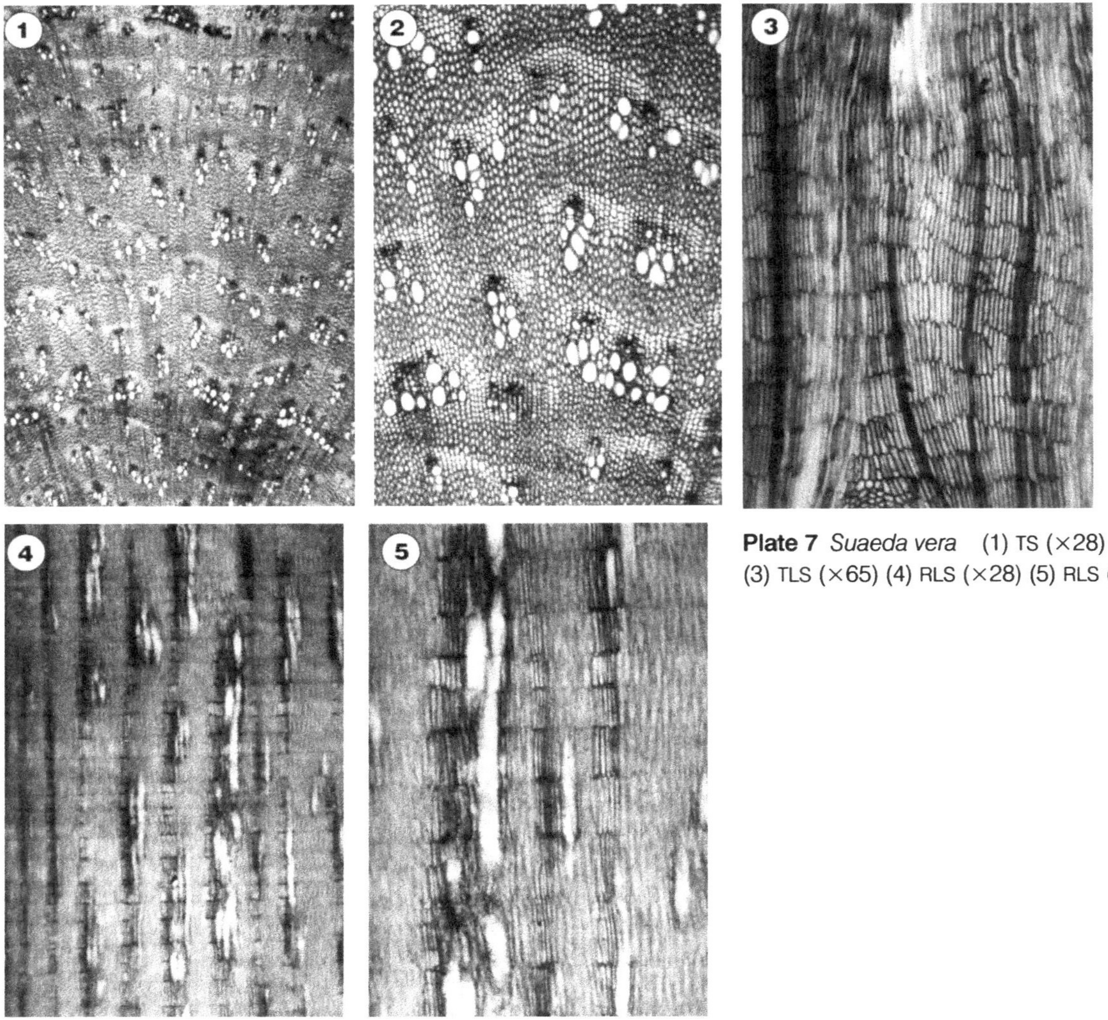

Plate 7 *Suaeda vera* (1) TS (×28) (2) TS (×65) (3) TLS (×65) (4) RLS (×28) (5) RLS (×65).

Group 3 – The ring porous woods

The ring porous wood group includes some of the most familiar of the larger trees, such as oak, ash and elm, commonly used in structural woodwork as well as in artefact manufacture. It also includes two commonly grown non-native fruit-bearing woody plants, mulberry and vine, as well as some less common shrubs.

Table 2.9 The morphology of the ring porous taxa

Quercus petraea, Q. robur	large trees
Ulmus glabra, U. minor sl, U. laevis	large trees
Fraxinus excelsior	large tree
Castanaea sativa	large tree
Morus nigra	small tree
Hippophae rhamnoides	large shrub
Berberis vulgaris	medium shrub
Rosa spp.	tree, shrub, scrambler
Clematis vitalba	liana
Vitis vinifera	vine

General anatomy

The ring porous woods are defined by having early wood vessels that are very much larger than the vessels in the late wood. In some cases tissues in the early and late wood are differently structured.

The ring porous nature of the wood of this group is very distinct and makes this group one of the easiest to identify. The ring of early wood vessels in most cases is so distinct that the late wood vessel grouping is treated entirely separately and used diagnostically (see Table 2.10). Perforation plates are simple throughout. Important characters are ray size visible in TS but more accurately seen in TLS, and the late wood vessel grouping seen in TS.

Table 2.10 Summary chart for the main characters used to differentiate the northern European ring porous woods

Taxa/character state	Late wood vessel grouping	Ray anatomy	Ray width	Ray height	Spiral thickening
Berberis vulgaris	Dendritic	Hm	(1) 4–8	–1 cm	present
Castanaea sativa	Dendritic	Hm (Ht 3)	1(2)	7–20	absent
Quercus spp.	Dendritic	Hm	1, 1 mm	7–20, –5 mm	absent
Rosa spp.	S	Ht 1	1, 6–15	3–12, –10 mm	fine
Hippophae rhamnoides	S, C	Hm (Ht 3)	1–2(–3)	5–15	present
Fraxinus excelsior	S, SRF, C	Ht 1,2,3	(1–)2–3	10–20	absent
Morus nigra	S, SRF, C	Ht 2,3	4–7	–60	present
Clematis vitalba	S, C	Ht 1,2,3	5–15+	10 cm+	present
Vitis vinifera	S, SRF (C)	Hm (Ht 2,3)	5–20	–2 mm	absent
Ulmus spp.	S, TC	Hm (Ht 3)	1, 4–5+	10–50	present

Note that in charcoal assemblages isolated fragments of the late wood of ring porous taxa can be found. These form from fracture planes that occur in the tangential plane through the early wood vessels. Often the tangential surfaces of such fragments have the remains of the early wood vessels upon them.

Taxa descriptions

Group key

Late wood pores in dendritic arrangement	*Quercus, Castanaea, Berberis*
Late wood pores mostly solitary	*Rosa*
Late wood pores solitary, in short radial files and/or clusters	
• Unisereate rays present	*Fraxinus, Hippophae*
• Unisereate rays absent	*Morus, Clematis, Vitis*
Late wood pores in tangential clusters	*Ulmus*

Late wood pores in dendritic arrangement

Quercus and *Castanea* (Fagaceae)

Quercus – the oaks (Fagaceae)

Plate 8

Based on the characters of the wood anatomy *Quercus* cannot be identified below the level of genus. The group includes *Q. robur*, *Q. patraea* and *Q. pubescens*. All three taxa have a native distribution (see Maps 10, 11 and 12).

Quercus is characterised by the ring porous organisation with a very distinct dendritic arrangement of late wood vessels. In narrow growth rings this may become compacted and in very wide growth rings the dendritic organisation can have a more radial orientation. Early wood vessels are often in one to three irregular rows but more commonly with the largest diameter vessels close to the initiation of the year's growth. Tyloses are common in the early wood vessels. Apotracheal parenchyma may be diffuse and in short tangential bands in the late wood. Paratracheal parenchyma is also sometimes present.

Rays are of two types, both homogeneous. Unisereate rays may be up to 20 cells high. Multisereate rays can be very large – up to 1 mm wide and 10 mm high. Multisereate rays are less frequent and smaller in immature wood. Perforation plates are simple. Vessel ray pits are large and slit-like.

There is possible confusion with *Berberis vulgaris* (see below) though *Quercus* has much wider and less regularly spaced multisereate rays and has much larger early wood vessels.

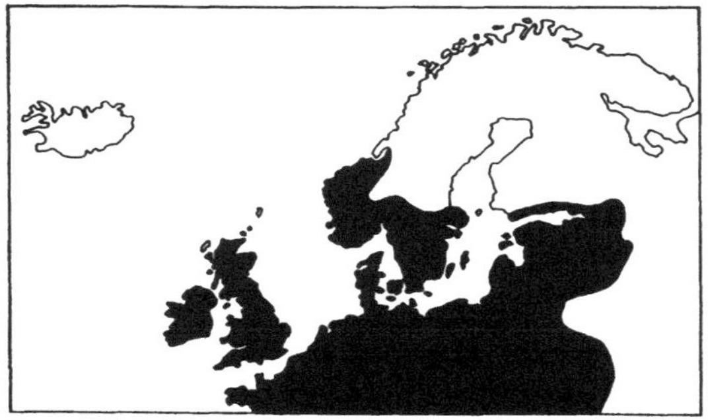

Map 10 Distribution of *Quercus robur*

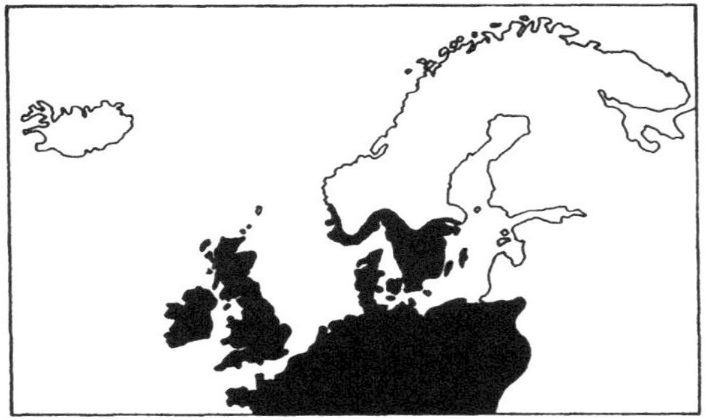

Map 11 Distribution of *Quercus patraea*

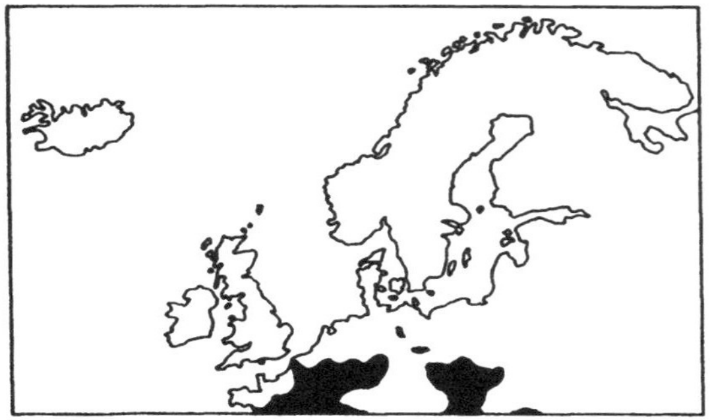

Map 12 Distribution of *Quercus pubescens*

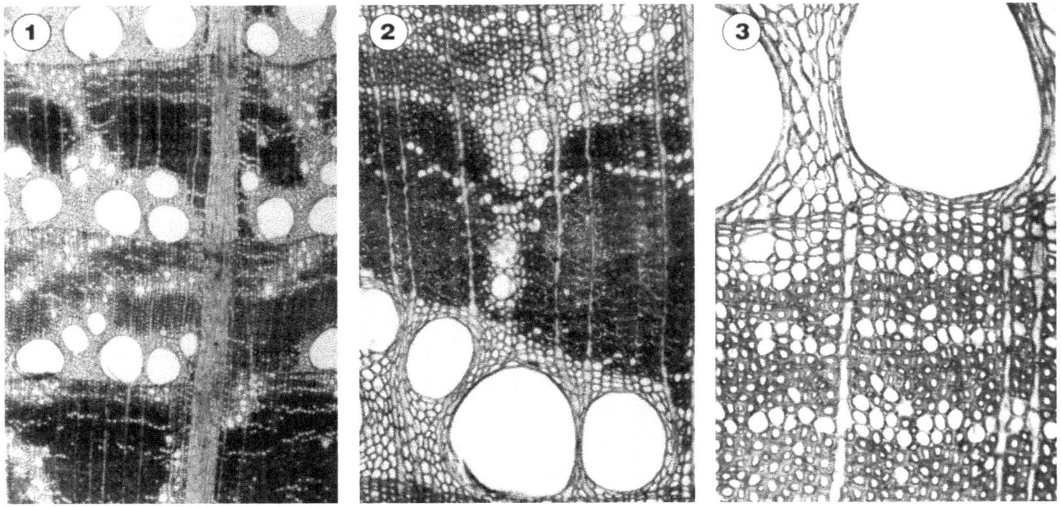

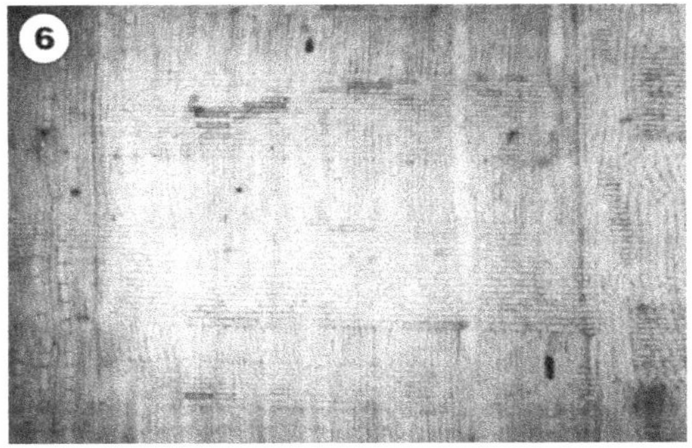

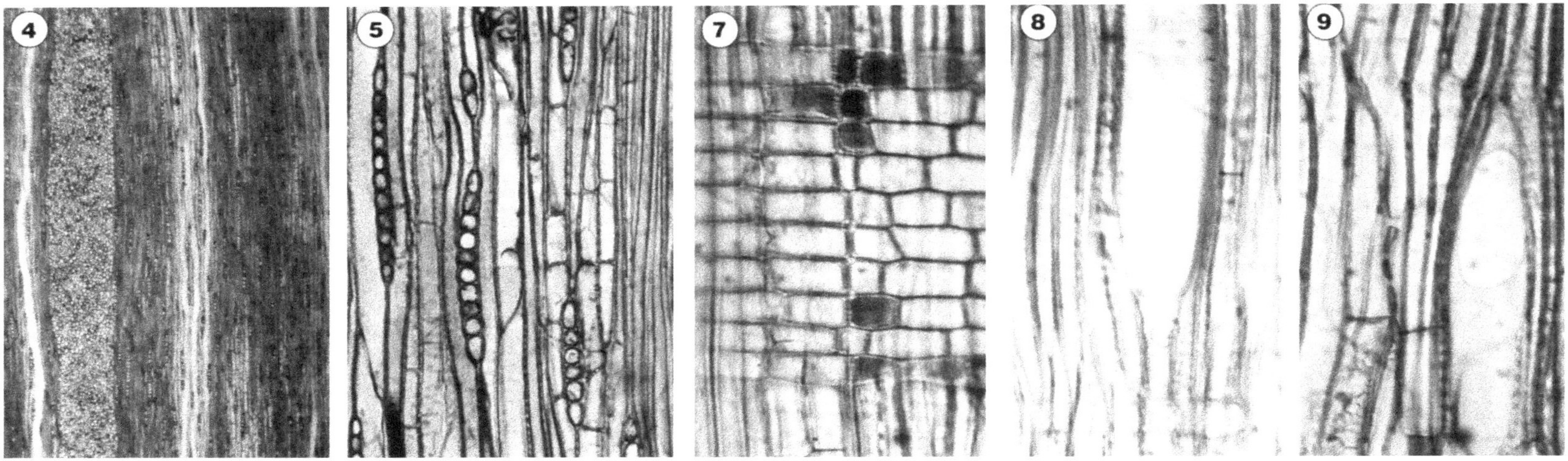

Plate 8 *Quercus robur* (1) TS (×28) (2) TS (×65) (3) TS showing tangential bands of parenchyma in late wood, (×140) (4) TLS (×28) (5) TLS (×140) (6) RLS (×28) (7) RLS (×140) (8) RLS simple perforation plate (×140) (9) RLS simple perforation plate (×140).

Castanea sativa – sweet chestnut (Fagaceae)

Plate 9

The anatomy of *Castanea sativa* is similar to *Quercus* in all respects (see above and Table 2.11) except in that of ray anatomy. In *Castanea* most of the rays are unisereate and homogeneous – the very large rays of *Quercus* are lacking. However some bisereate and even rare trisereate rays occur and occasional rows of square cells may be seen in rays in RLS. Schweingruber (1990) notes that scalariform perforation plates may be present in the late wood vessels.

There is possible confusion with immature *Quercus* where the large multisereate rays, typical of mature wood, are reduced in size and number.

Castanea is most likely to have been introduced to southern areas of northern Europe in the Roman period, although it is possible that earlier isolated late prehistoric introductions occurred. Its natural distribution is in the mid and west Mediterranean, however see Map 13 for its likely post-Roman distribution.

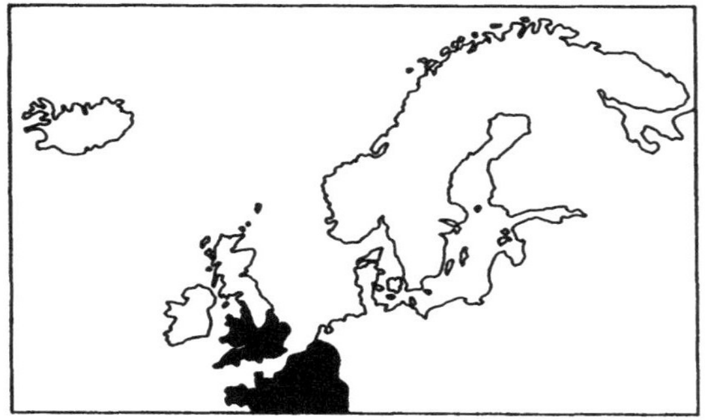

Map 13 Distribution of *Castanaea sativa*

Table 2.11 Anatomical characters of *Quercus* and *Castanea sativa* (Fagaceae)

Character	*Castanea sativa*	*Quercus*
TS		
Vessel arrangement	ring porous	ring porous
Vessel grouping	dendritic	dendritic
Early–late wood transition	abrupt	abrupt
Late wood vessels	thin-walled	thin-walled
Tyloses	present in early wood	present in early wood
Parenchyma	apotracheal diffuse/ tangential rows (paratracheal)	apotracheal diffuse/ tangential rows (paratracheal)
Ground tissue	–	–
Ring boundary	–	–
Ray/ring boundary	–	–
TLS		

Character	*Castanea sativa*	*Quercus*
Ray width	1(2)	1, –1 mm
Ray height	10–20	10–20, –10 mm
Sheath cells	absent	absent
RLS		
Ray anatomy	homogeneous (heterogeneous)	homogeneous
Libriform fibres	present	present
Fibre tracheids	absent	absent
Vasicentric tracheids	present	present
Vascular tracheids	absent	absent
Spiral thickening	absent	absent
Crystals	(present)	present
Perforation plates	simple	simple

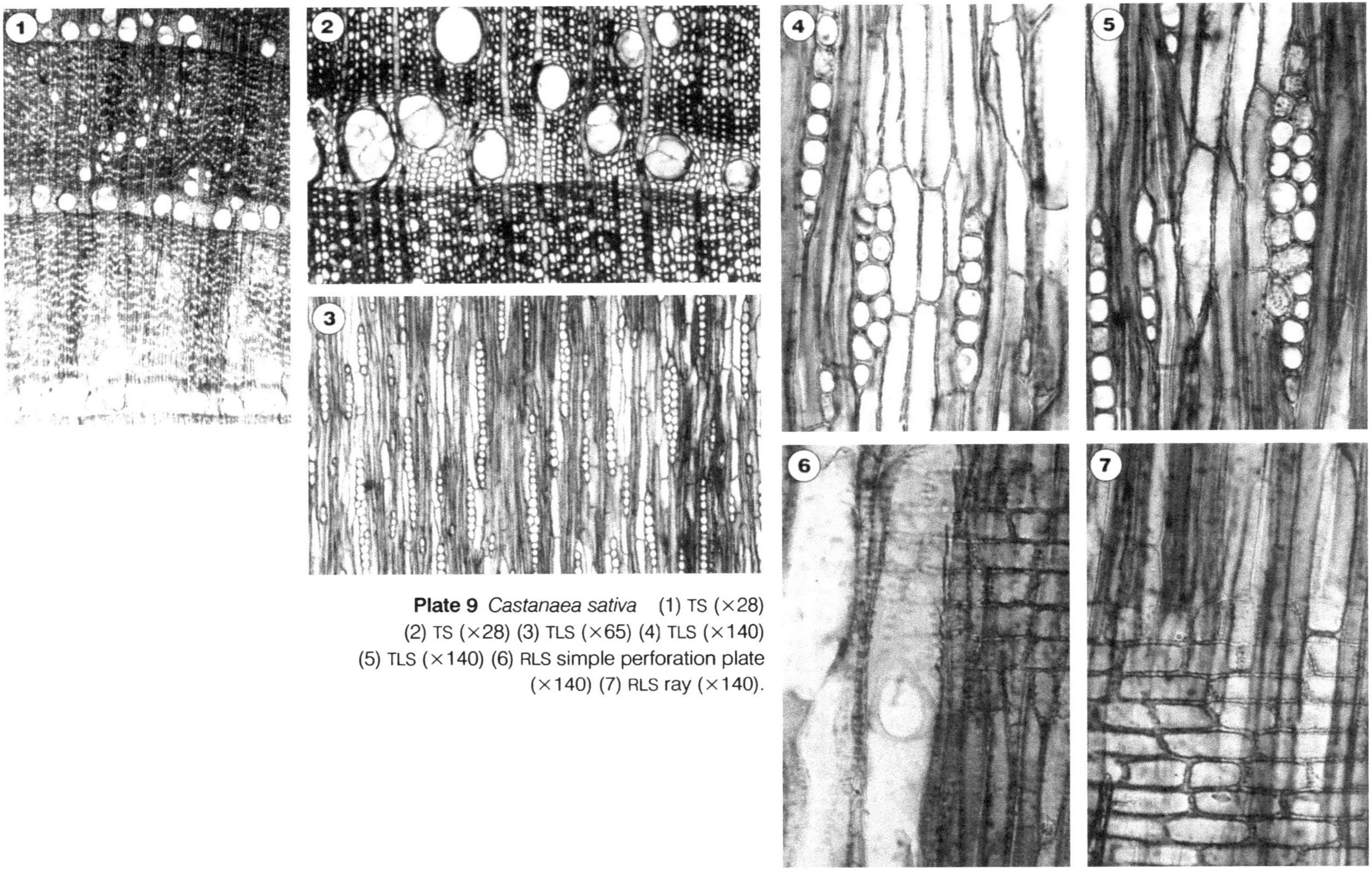

Plate 9 *Castanaea sativa* (1) TS (×28) (2) TS (×28) (3) TLS (×65) (4) TLS (×140) (5) TLS (×140) (6) RLS simple perforation plate (×140) (7) RLS ray (×140).

Berberis vulgaris – Barberry (Berberidaceae)

Plate 10

Berberis vulgaris is a small thorny shrub of central and western northern Europe, naturalised elsewhere (see Map 14). Its status in the United Kingdom is questionable but most likely non-native. Its wood is distinctively yellow.

Berberis vulgaris has a strongly undulating ring boundary and a weak, often discontinuous, ring of early wood vessels that may appear to reduce in size close to the initiation of the late wood such that there is a continuous gradation. Late wood vessels are arranged dendritically and appear to account for much of the ring's width. Parenchyma is apotracheal diffuse.

Rays are homogeneous, regularly spaced, 4–8 cells wide and up to 1 cm high, often appearing in TLS to be very high and narrow. In TS they appear slightly distended at the ring boundary. Very much smaller unisereate and bisereate rays are infrequent, perforation plates are simple.

There is possible confusion with *Quercus*, though the multisereate rays are much narrower and more regularly spaced, and the early wood vessels are much smaller.

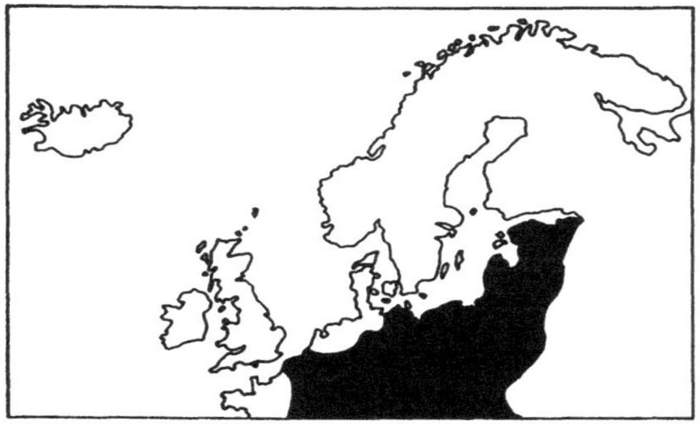

Map 14 Distribution of *Berberis vulgaris*

Table 2.12 Anatomical characters of *Berberis vulgaris* (Berberidaceae)

Character	*Berberis vulgaris*	Character	*Berberis vulgaris*
TS		RLS	
Vessel arrangement	ring porous to semi-ring porous	Ray anatomy	homogeneous
Vessel grouping	solitary and concentric – dendritic	Libriform fibres	present
Tyloses	absent	Fibre tracheids	absent
Parenchyma	apotraceal diffuse	Vascular tracheids	absent
Ground tissue	–	Spiral thickening	present
Ring boundary	undulating	Perforation plates	simple
Ray/ring boundary	rays distended		
TLS			
Ray width	(1–2) 4–8		
Ray height	–1 cm		
Sheath cells	(present)		

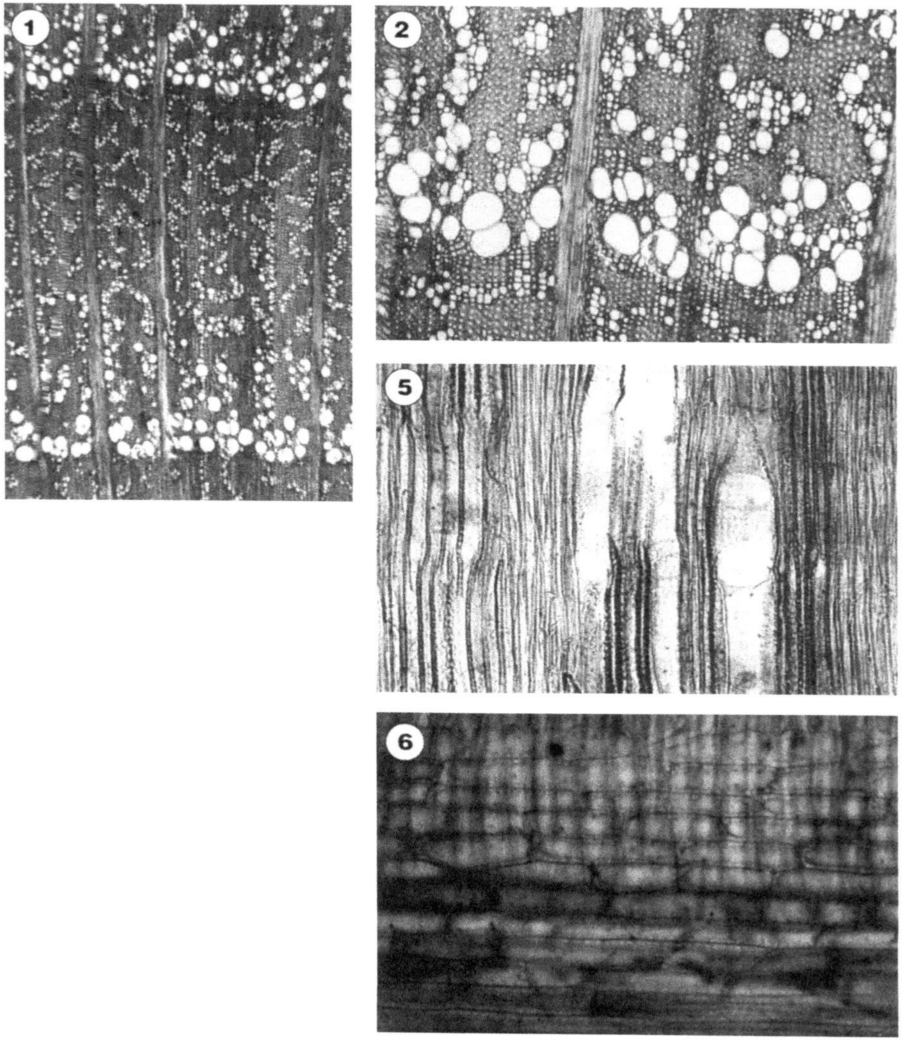

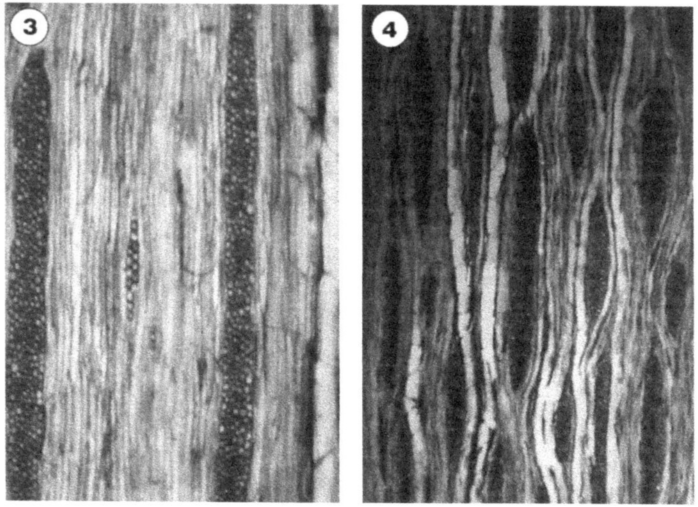

Plate 10 *Berberis vulgaris* (1) TS (×28) (2) TS (×65) (3) TLS (×65) (4) TLS (×65) (5) RLS simple perforation plate (×65) (6) RLS ray (×140).

Late wood pores mostly solitary

Rosa – the roses (Rosaceae – Rosoideae)

Plate 11

This genus is made up of many species that cannot be separated anatomically. The commoner species are listed here with their distributions. Roses may be small or large shrubs, small trees, lianas or scramblers.

Rosa obtusifolia (Map 15)
Rosa caesia (Map 16)
Rosa stylosa (Map 17)
Rosa cannina (Map 18)
Rosa tormentosa (Map 19)
Rosa sherardii (Map 20)
Rosa mollis (Map 21)
Rosa agristris (Map 22)
Rosa arvensis (Map 23)
Rosa pimpinellifolia (Map 24)
Rosa micrantha (Map 25)
Rosa rubininosa (Map 26)

Rosa is mostly ring porous with a tendency towards the semi-ring porous state. The early wood vessels are never large but form a discernible irregularly continuous ring of one to three vessels. Late wood vessels are mostly solitary with very few clusters of two to three vessels, ground tissue is thick-walled. Apotracheal parenchyma is diffuse and paratracheal parenchyma is rare.

Multisereate rays are distended at the ring boundary (seen in TS), up to 15 cells wide and often over 10 mm high. Sheath cells are common (seen in TLS). The rays are heterogeneous with many rows of upright cells. Unisereate rays are up to 12 cells high and made up entirely of upright cells. Spiral thickening is fine but universally present. Perforation plates are simple.

There is possible confusion with ring porous woods that have late wood vessels clusters or radial files as well as solitary vessels. Three of these *Morus*, *Vitis* and *Clematis* have large multisereate rays but no unisereate rays. In all three the late wood vessel distribution is also distinctly different to *Rosa*, having far fewer solitary vessels. In the two ring porous taxa with unisereate rays and solitary vessels as well as clusters in the late wood, *Fraxinus* and *Hippophae*, both lack large multisereate rays.

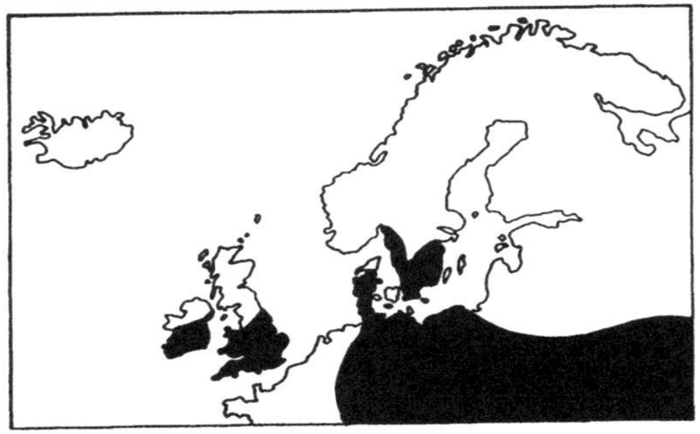

Map 15 Distribution of *Rosa obtusifolia*

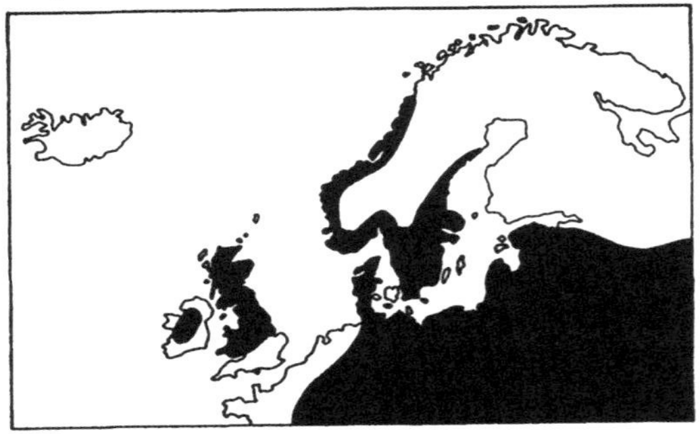

Map 16 Distribution of *Rosa caesia*

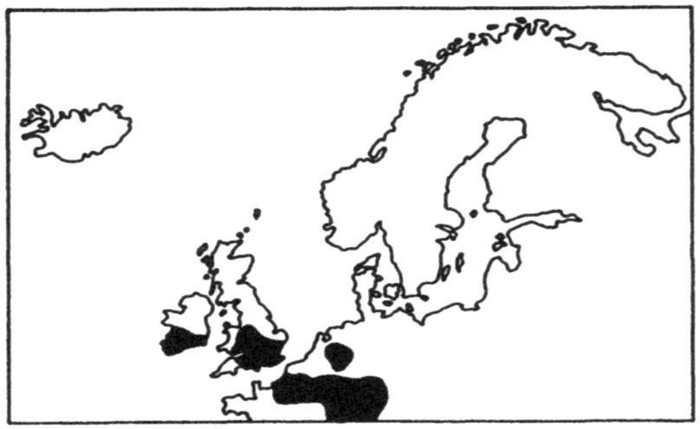

Map 17 Distribution of *Rosa stylosa*

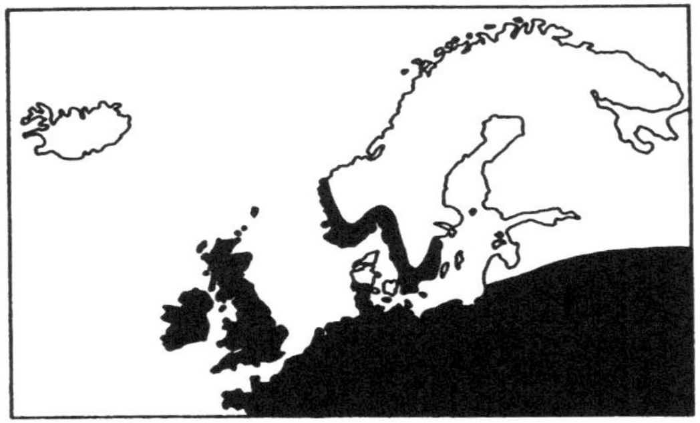

Map 18 Distribution of *Rosa cannina*

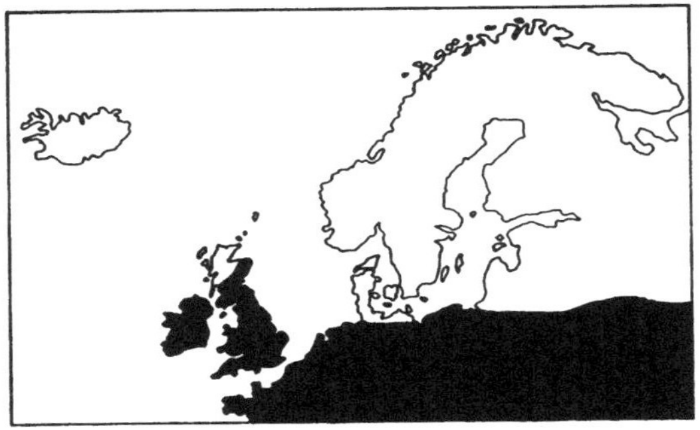

Map 19 Distribution of *Rosa tormentosa*

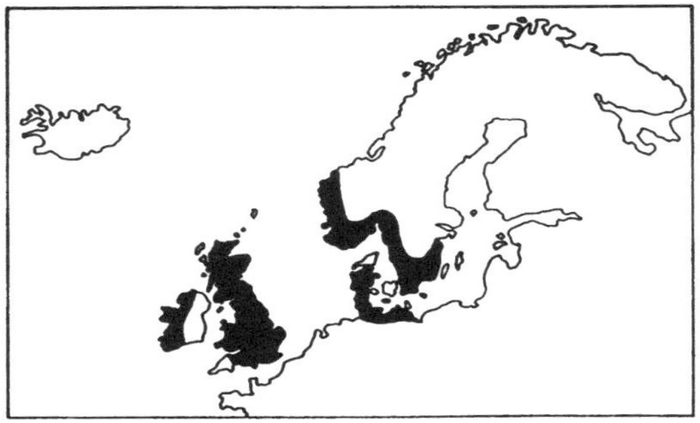

Map 20 Distribution of *Rosa sherardii*

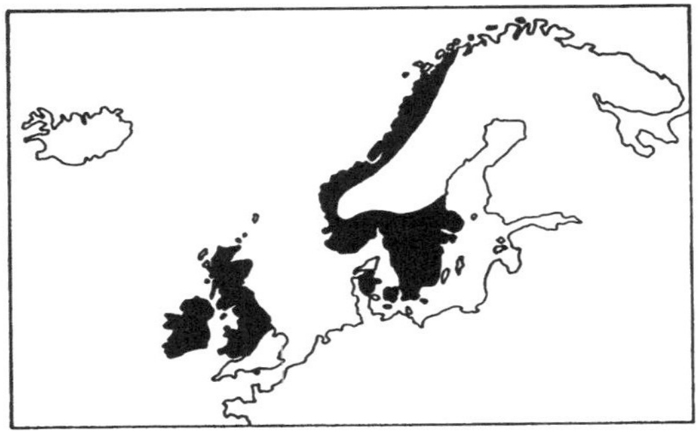

Map 21 Distribution of *Rosa mollis*

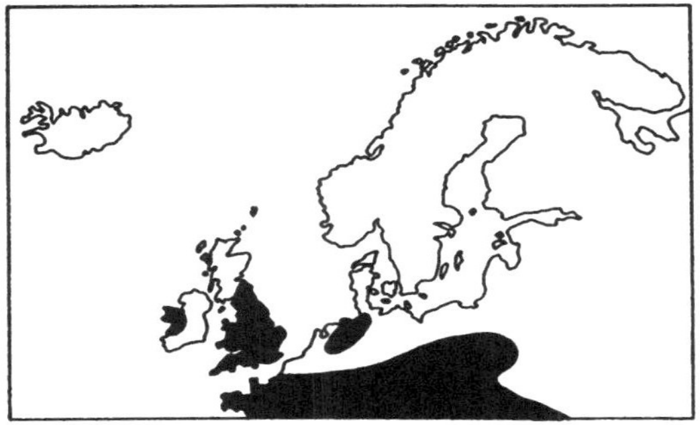

Map 22 Distribution of *Rosa agristris*

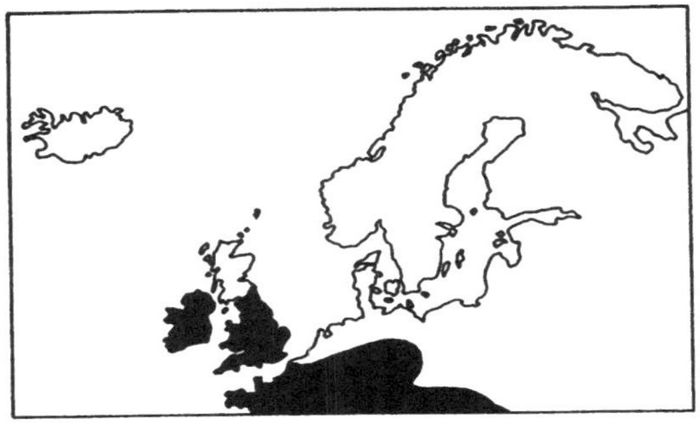

Map 23 Distribution of *Rosa arvensis*

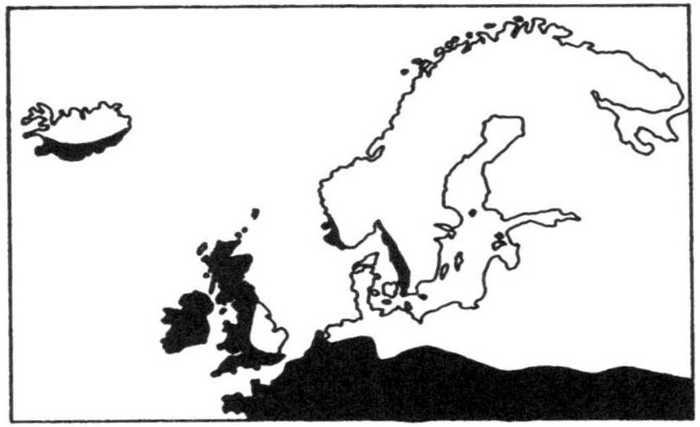

Map 24 Distribution of *Rosa pimpinellifolia*

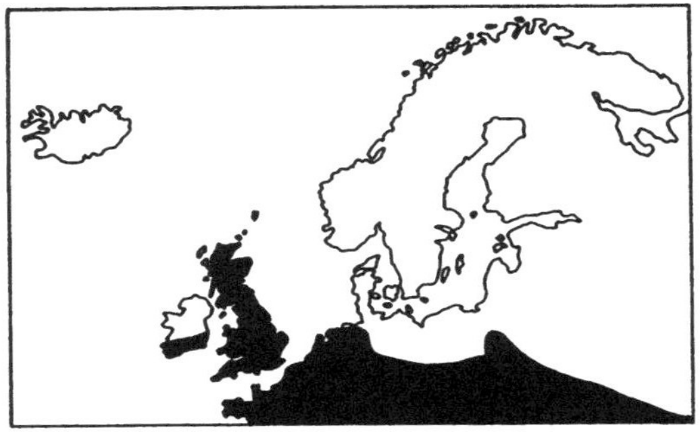

Map 25 Distribution of *Rosa micrantha*

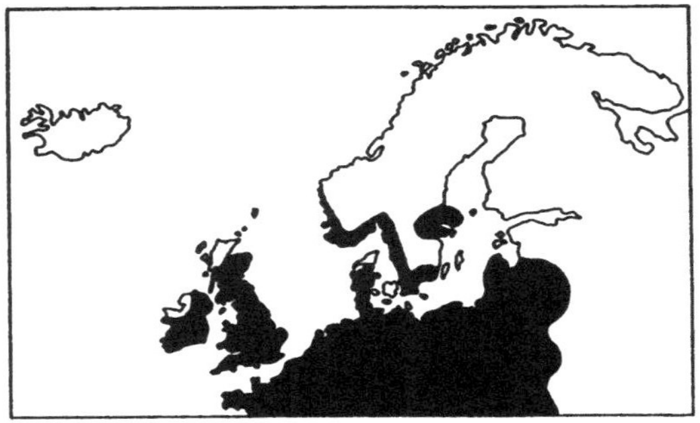

Map 26 Distribution of *Rosa rubininosa*

Table 2.13 Anatomical characters of *Rosa* spp (Rosaceae – Rosoideae)

Character	*Rosa* spp
TS	
Vessel arrangement	ring porous (to semi-ring porous)
Vessel grouping	solitary
Tyloses	absent
Parenchyma	Apotracheal diffuse (paratracheal)
Ground tissue	thick-walled
Ring boundary	distinct
Ray/ring boundary	(rays distended)
TLS	
Ray width	1, 6–15
Ray height	3–12, –10 mm
Sheath cells	present
RLS	
Ray anatomy	heterogeneous type 1
Libriform fibres	absent
Fibre tracheids	present
Vascular tracheids	absent
Spiral thickening	fine
Crystals	present
Perforation plates	simple

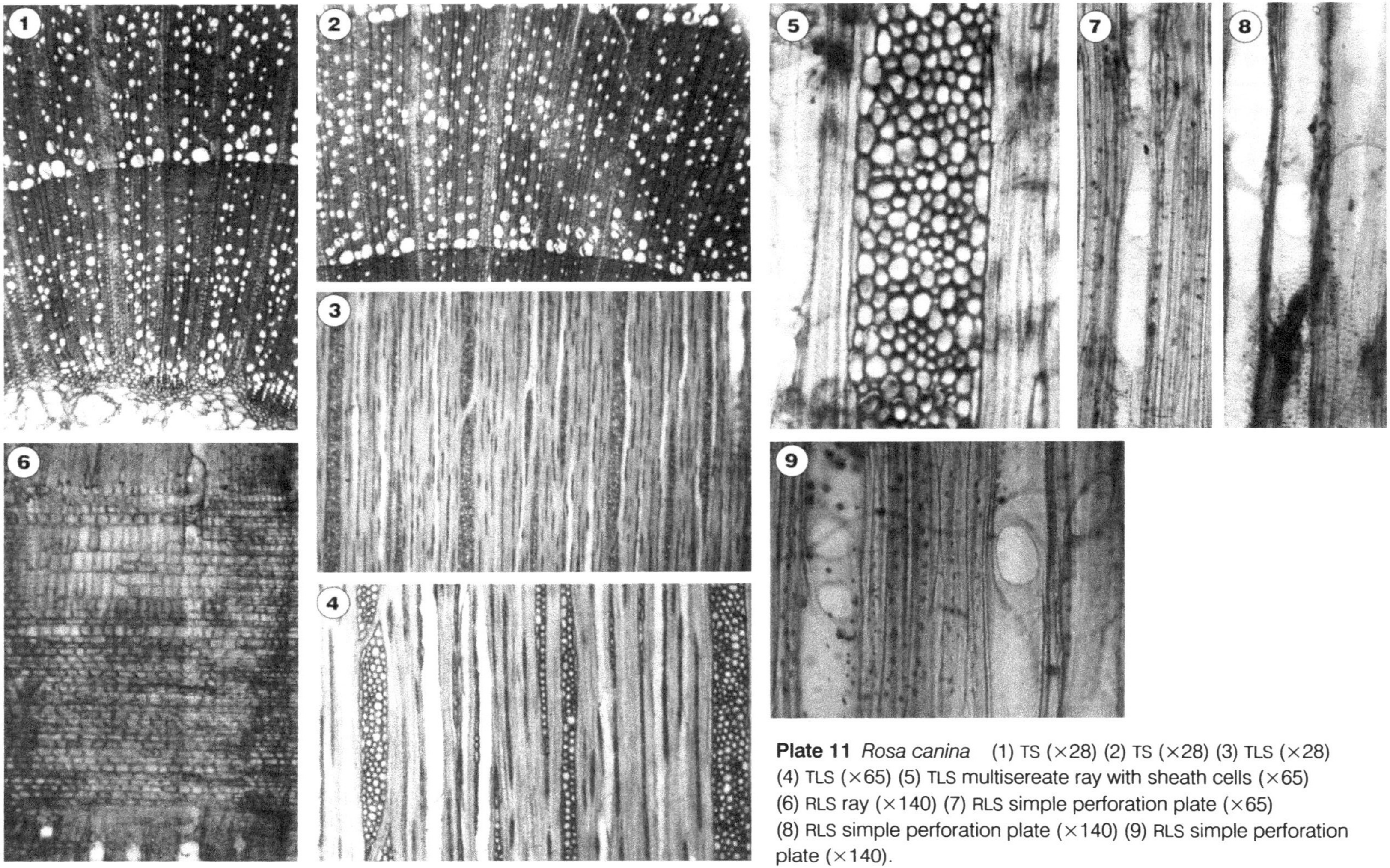

Plate 11 *Rosa canina* (1) TS (×28) (2) TS (×28) (3) TLS (×28) (4) TLS (×65) (5) TLS multisereate ray with sheath cells (×65) (6) RLS ray (×140) (7) RLS simple perforation plate (×65) (8) RLS simple perforation plate (×140) (9) RLS simple perforation plate (×140).

Late wood pores solitary, in short radial files and/or clusters. Unisereate rays present

Fraxinus excelsior – Ash (Oleaceae)

Plate 12

Fraxinus is an important large woodland tree as well as being a common structural and artefactual wood. It has a wide distribution across northern Europe (see Map 27).

The anatomy of *Fraxinus* is distinct from other ring porous woods. Early wood vessels are large and solitary or in short radial files. Late wood vessels are solitary, in small clusters or in short radial files. Parenchyma is distinctly paratracheal vasicentric and is very visible in LS as well as in TS. Terminal parenchyma also present. Ground tissue becomes thick-walled in late wood. Rays are bi- to trisereate, occasionally unisereate, up to 20 cells high and mostly homogeneous with occasional rows of square cells. Perforation plates are simple.

There is possible confusion with *Hippophae* (below) though this does not have the very distinctive paratracheal parenchyma.

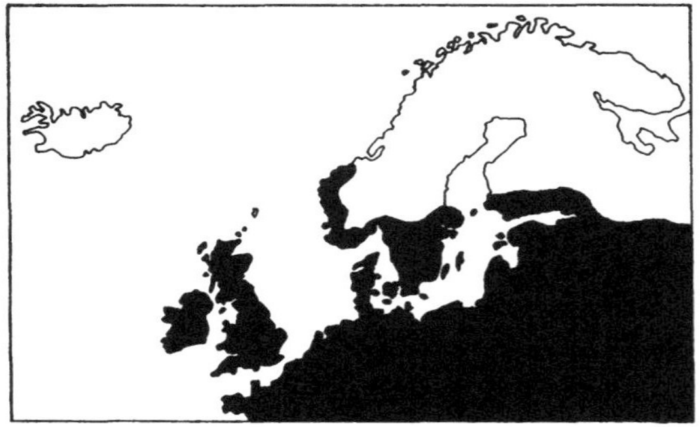

Map 27 Distribution of *Fraxinus excelsior*

Table 2.14 Anatomical characters of *Fraxinus excelsior* (Oleaceae)

Character	*Fraxinus excelsior*	Character	*Fraxinus excelsior*
TS		RLS	
Vessel arrangement	ring porous	Ray anatomy	homogeneous (heterogeneous type 3)
Vessel grouping	solitary, short radial files, clusters	Libriform fibres	present
Tyloses	absent	Fibre tracheids	absent
Parenchyma	paratracheal vasicentric, terminal	Vascular tracheids	absent
Ground tissue	thick-walled in late wood	Spiral thickening	absent
Ring boundary	distinct	Crystals	present
Ray/ring boundary	–	Perforation plates	simple
TLS			
Ray width	(1–) 2–3		
Ray height	10–20		
Sheath cells	absent		

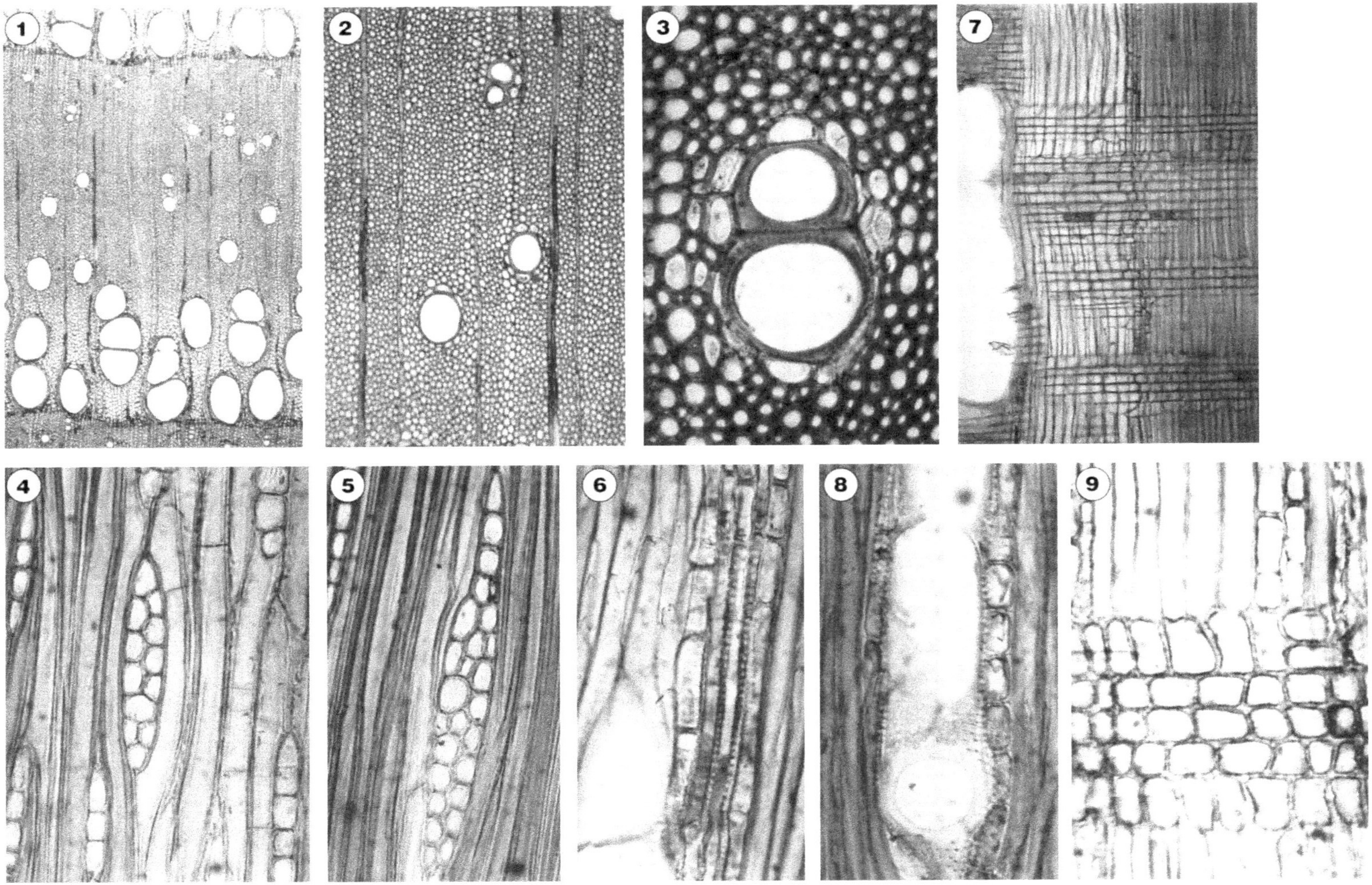

Plate 12 *Fraxinus excelsior* (1) TS (×28) (2) TS late wood vessels with paratracheal parenchyma (×65) (3) TS late wood vessel with paratracheal parenchyma (×265) (4) TLS (×265) (5) TLS (×265) (6) TLS paratracheal parenchyma (×265) (7) RLS ray (×140) (8) RLS simple perforation plate (×265) (9) RLS simple perforation plate (×265).

Hippophae rhamnoides – Sea Buckthorn (Eleagnaceae)

Plate 13

Hippophae rhamnoides is a large thorny shrub of coastal margins found on northern European and Scandinavian coasts as well as the east coast of the United Kingdom (see Map 28).

The large early wood vessels of Hippophae are in many rows and may be solitary, in clusters or short radial files. The vessels are often oval in shape with the long axis radially positioned. Also the largest of the early wood vessels are away from the ring boundary. Those adjacent to the ring boundary are somewhat smaller. Late wood vessels are solitary and in small irregular clusters. Rays are unisereate to bisereate rarely larger, up to 15 cells high, mostly homogeneous and often storied. Perforation plates are simple.

There is possible confusion with *Fraxinus* (above) though *Hippophae* does not have distinctly paratracheal parenchyma.

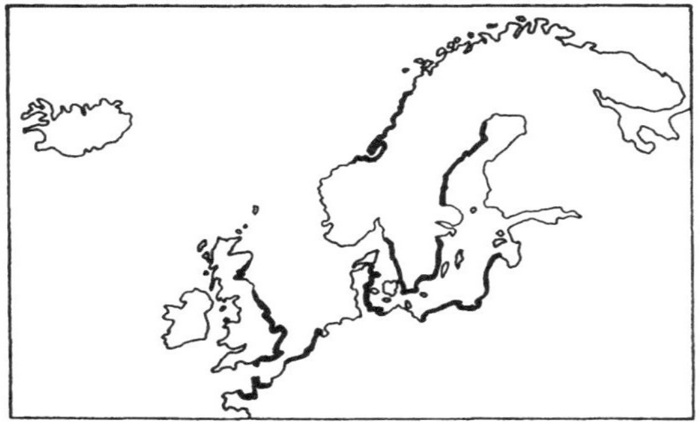

Map 28 Distribution of *Hippophae rhamnoides*

Table 2.15 Anatomical characters of *Hippophae rhamnoides* (Eleagnaceae)

Character	*Hippophae rhamnoides*	Character	*Hippophae rhamnoides*
TS		RLS	
Vessel arrangement	ring porous	Ray anatomy	homogeneous (heterogeneous type 3)
Vessel grouping	solitary, clusters	Libriform fibres	absent
Tyloses	present	Fibre tracheids	present
Parenchyma	apotracheal, diffuse (paratracheal)	Vascular tracheids	absent
Ground tissue	thin-walled in early wood	Spiral thickening	present
Ring boundary	distinct	Perforation plates	simple
Ray/ring boundary	–		
TLS			
Ray width	1–2(–3)		
Ray height	5–15		
Sheath cells	absent		

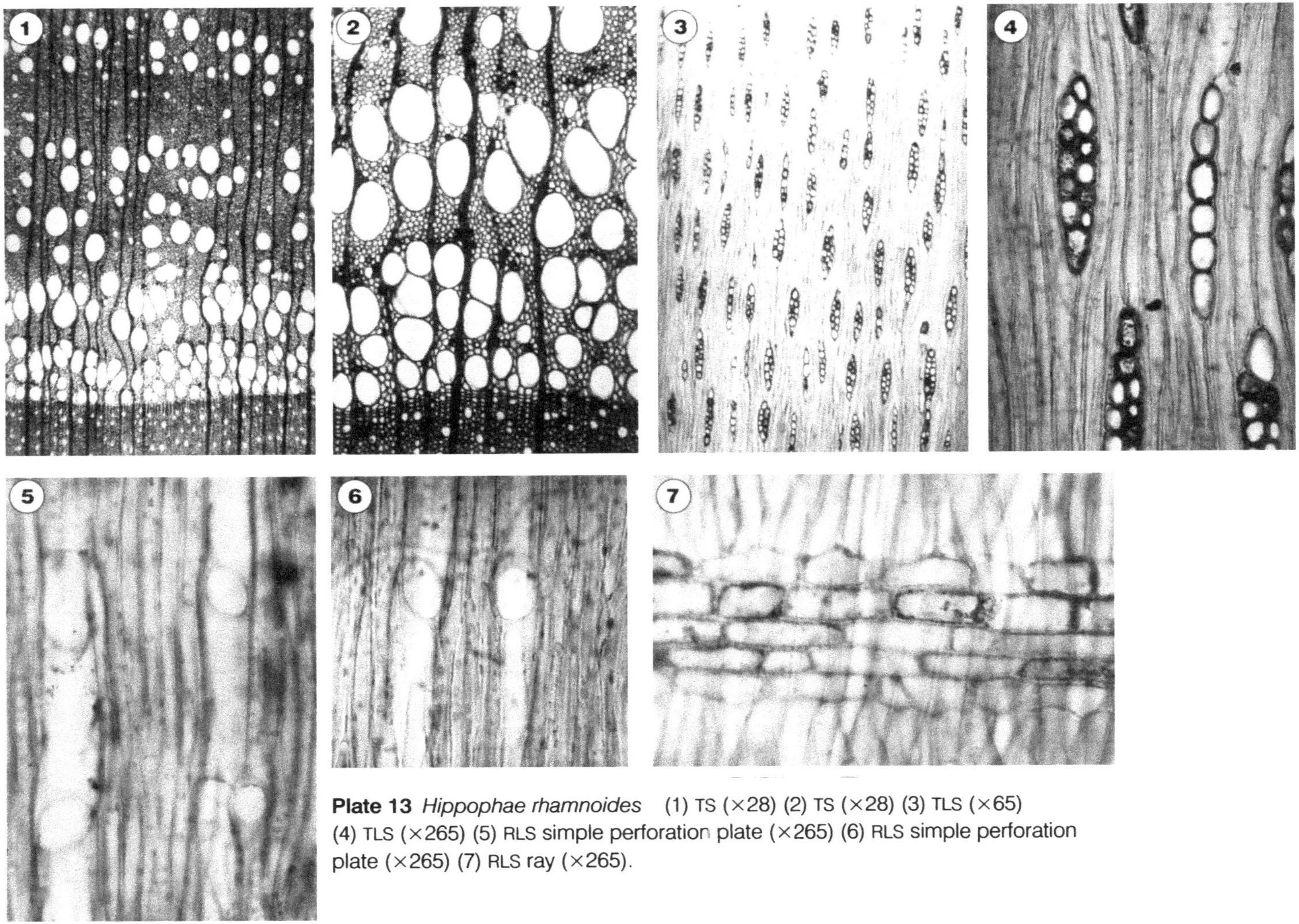

Plate 13 *Hippophae rhamnoides* (1) TS (×28) (2) TS (×28) (3) TLS (×65) (4) TLS (×265) (5) RLS simple perforation plate (×265) (6) RLS simple perforation plate (×265) (7) RLS ray (×265).

Late wood pores in short radial files and or clusters. Unisereate rays absent

Morus nigra – Black mulberry (Moraceae)

Plate 14

Morus nigra is a late medieval introduction into Europe from central Asia. It is a small tree cultivated for its fruit though is now naturalised in places.

This species is quite variable in the size and number of rows of early wood vessels, which may be solitary, in clusters or short radial files. Late wood vessels are in clusters with paratracheal parenchyma, vascular tracheids and thick-walled ground tissue. Rays are multisereate, generally 4 to 7 cells wide but often larger and up to 60 cells high. Sheath cells are often present. Rays are homogeneous. Perforation plates are simple.

This taxon may be confused with *Rosa* (above) when there are few early wood vessels, though *Morus* has no unisereate rays, or *Ulmus* (below) when early wood vessels are present as one or two continuous rows. The clusters of late wood vessels are weaker than in *Ulmus* and rays are higher.

Table 2.16 Anatomical characters of *Morus nigra* (Moraceae)

Character	*Morus nigra*	Character	*Morus nigra*
TS		RLS	
Vessel arrangement	ring porous	Ray anatomy	heterogeneous types 2, 3
Vessel grouping	solitary, clusters, short radial files	Libriform fibres	present
Tyloses	present	Fibre tracheids	absent
Parenchyma	paratracheal	Vascular tracheids	present
Ground tissue	–	Spiral thickening	present
Ring boundary	–	Crystals	present
Ray/Ring boundary	–	Perforation plates	simple
TLS			
Ray width	–4–7(7+)		
Ray height	–60		
Sheath cells	occasional		

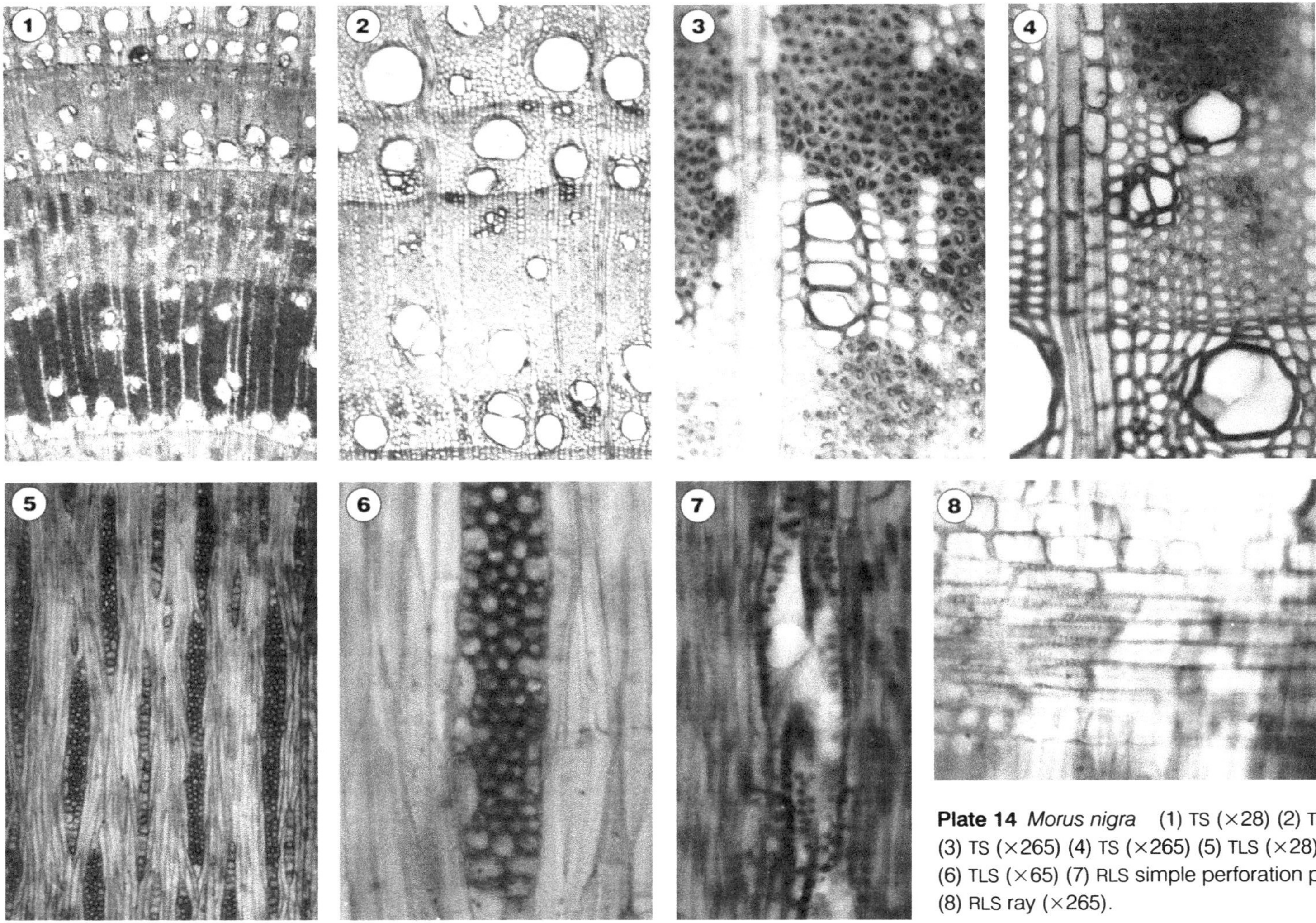

Plate 14 *Morus nigra* (1) TS (×28) (2) TS (×65) (3) TS (×265) (4) TS (×265) (5) TLS (×28) (6) TLS (×65) (7) RLS simple perforation plate (×265) (8) RLS ray (×265).

Clematis vitalba – Traveller's joy (Ranunculaceae)

Plate 15

Clematis vitalba is a liana – growing through the crown of trees in a woodland or hedgerow. Its distribution in northern Europe is distinctly western (see Map 29).

Clematis vitalba has a very distinct ring porous vessel distribution with very large, solitary early wood vessels and a narrow band of late wood vessels that appear in irregular often tangentially oriented clusters. Growth ring boundaries are concave towards the pith between the rays, but are generally rather indistinct. Rays are up to 15 cells wide and may be over 1 cm high, heterogeneous and with sheath cells. Perforation plates are simple.

It would be difficult to confuse *Clematis* with any other wood, except perhaps *Vitis* (below) though this has very distinct late wood radial pore files rather than the irregular tangential clusters of *Clematis*.

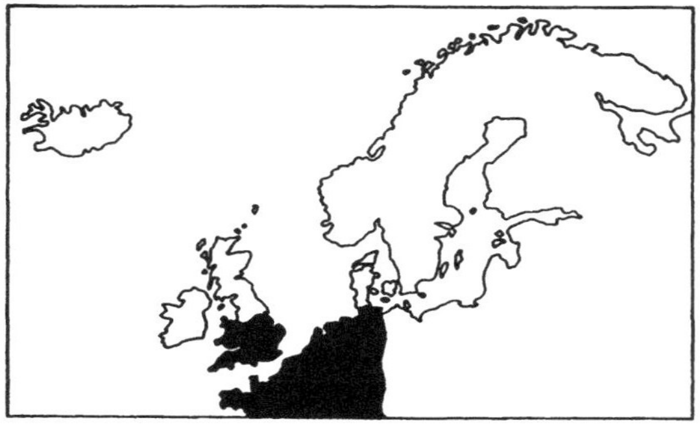

Map 29 Distribution of *Clematis vitalba*

Table 2.17 Anatomical characters of *Clematis vitalba* (Ranunculaceae)

Character	*Clematis vitalba*	Character	*Clematis vitalba*
TS		RLS	
Vessel arrangement	ring porous	Ray anatomy	heterogeneous types 1, 2, 3
Vessel grouping	solitary, clusters	Libriform fibres	present
Tyloses	absent	Fibre tracheids	absent
Parenchyma	apotracheal, paratracheal	Vascular tracheids	present
Ground tissue	–	Spiral thickening	present
Ring boundary	indistinct, undulating	Crystals	absent
Ray/ring boundary	–	Perforation plates	simple
TLS			
Ray width	5–15+		
Ray height	1 cm+		
Sheath cells	present		

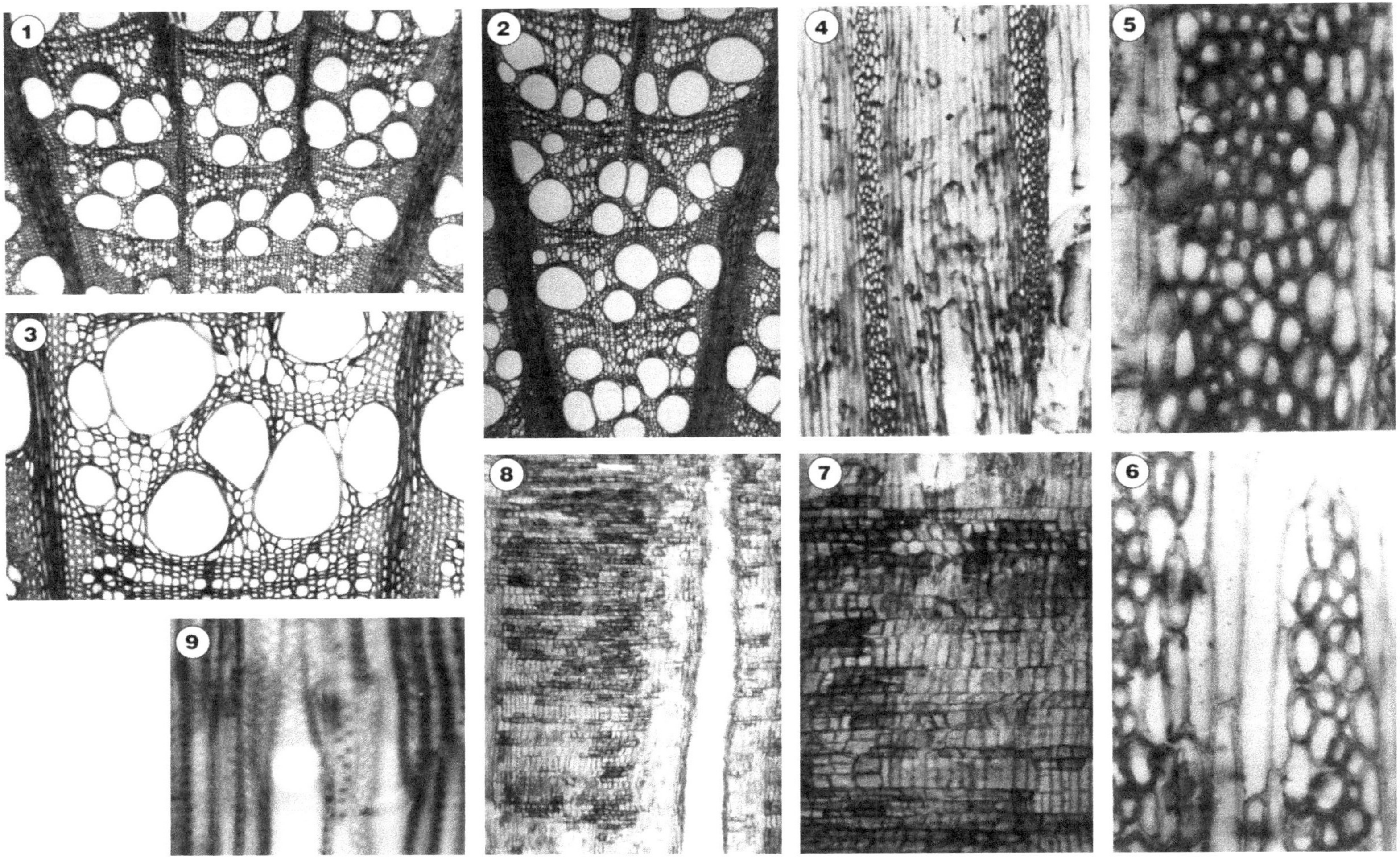

Plate 15 *Clematis vitalba* (1) TS (×28) (2) TS (×28) (3) TS (×65) (4) TLS (×65) (5) TLS multisereate ray with sheath cells (×265) (6) TLS multisereate ray with sheath cells (×265) (7) RLS ray (×140) (8) RLS ray (×65) (9) RLS simple perforation plate (×140)

Vitis vinifera – Grape vine (Vitaceae)

Plate 16

Vitis is probably not native to northern Europe and is an introduction from the Mediterranean region certainly in Roman times and probably before, cultivated for its fruit. It can grow to be a substantial vine with a lower stem of considerable thickness.

Vitis vinifera has a ring porous distribution with very large oval-shaped early wood vessels with their long axes radially positioned. Late wood vessels are in long radial files. The growth ring is gently undulating. Rays are wide and closely spaced such that generally there is only one early wood vessel between each ray. Rays are generally up to 20 cells wide and 2 mm high. Homogeneous and heterogeneous rays are present and sheath cells are present. Intervessel pits are distinctly scalariform, perforation plates are simple.

Confusion of this species may only arise with *Clematis* though the late wood distribution of vessels is distinctly different.

Table 2.18 Anatomical characters of *Vitis vinifera* (Vitaceae)

Character	*Vitis vinifera*	Character	*Vitis vinifera*
TS		RLS	
Vessel arrangement	ring porous (semi- ring porous)	Ray anatomy	homogeneous, heterogeneous types 2, 3
Vessel grouping	solitary, short radial files (clusters)	Libriform fibres	present (septate)
Tyloses	present	Fibre tracheids	absent
Parenchyma	paratracheal	Vascular tracheids	present
Ground tissue	–	Spiral thickening	absent
Ring boundary	distinct undulating	Crystals	present
Ray/ring boundary	–	Perforation plates	simple (scalariform)
TLS			
Ray width	5–20		
Ray height	–2mm		
Sheath cells	(present)		

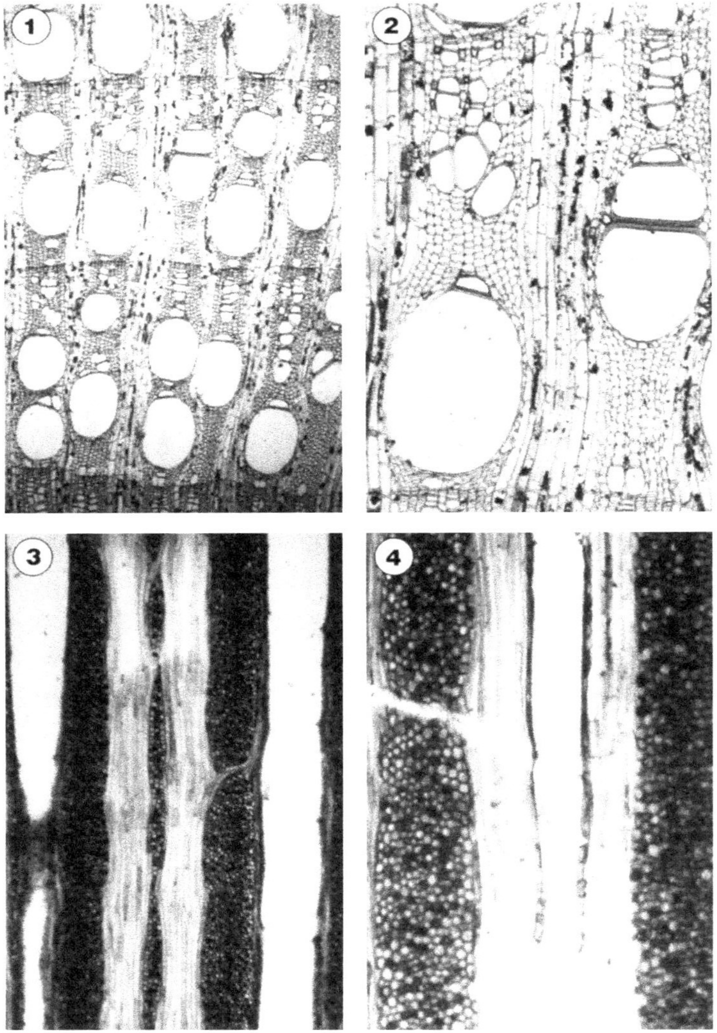

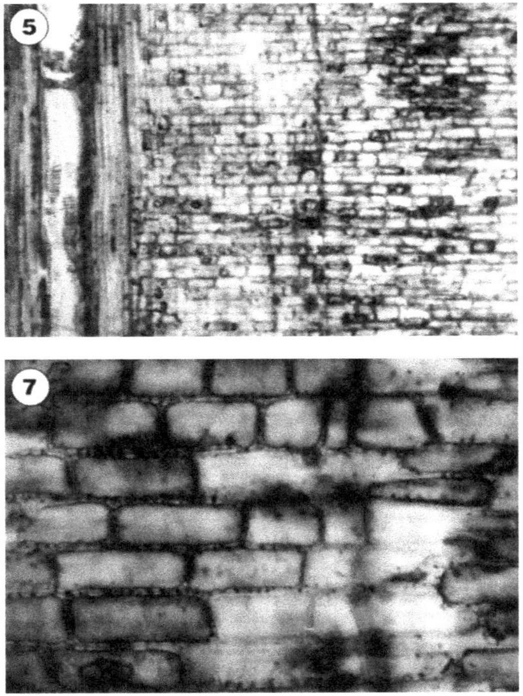

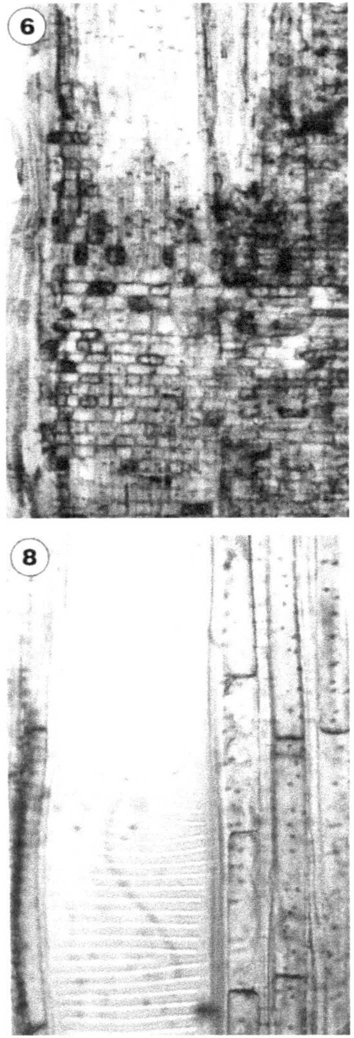

Plate 16 *Vitis vinifera* (1) TS (×28) (2) TS (×140) (3) TLS (×65) (4) TLS (×140) (5) RLS ray (×140) (6) RLS ray (×140) (7) RLS ray (×265) (8) RLS simple perforation plate and scalariform pitting (×265).

Late wood pores in tangential clusters

Ulmus – the elms (Ulmaceae)

Plate 17

The three northern European species of *Ulmus* are all large woodland and hedgerow trees with broad distributions. They cannot be separated by their wood anatomy.

Ulmus glabra – Wych elm (see Map 30)
Ulmus laevis – Flattering elm (see Map 31)
Ulmus minor agg. – smooth-leaved elm / English elm (see Map 32)

The anatomy of *Ulmus* is ring porous but with very diagnostic late wood distribution of vessels. There is usually only one row of large rounded early wood vessels, though there may be up to three. The late wood vessels are in gently undulating tangential rows together with vascular tracheids and paratracheal parenchyma. Ground tissue is thick-walled. Rays are distended at the ring boundary, 4 to 5 cells wide and up to 50 cells high. Small unisereate rays are not infrequent. Rays may be homogeneous or heterogeneous. Perforation plates are simple.

Ulmus cannot easily be confused with other ring porous woods though an isolated fragment of the late wood may be confused with *Hedera helix* that has similar tangential bands of vessels and a thick-walled ground tissue. However *Hedera* has much larger multisereate rays.

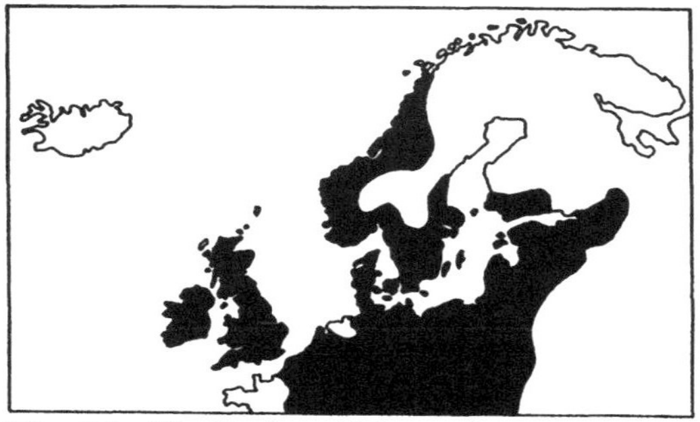

Map 30 Distribution of *Ulmus glabra*

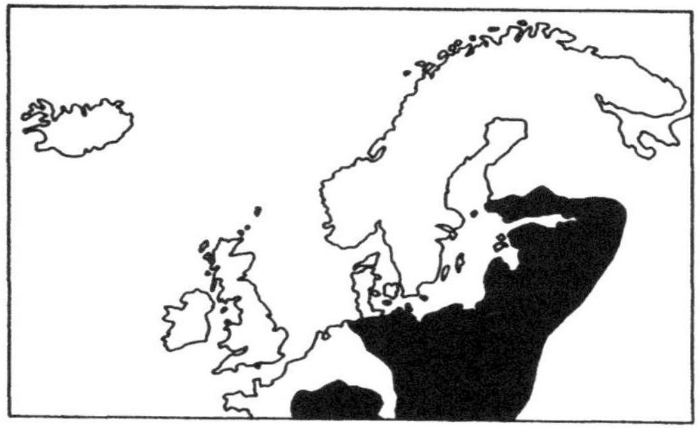

Map 31 Distribution of *Ulmus laevis*

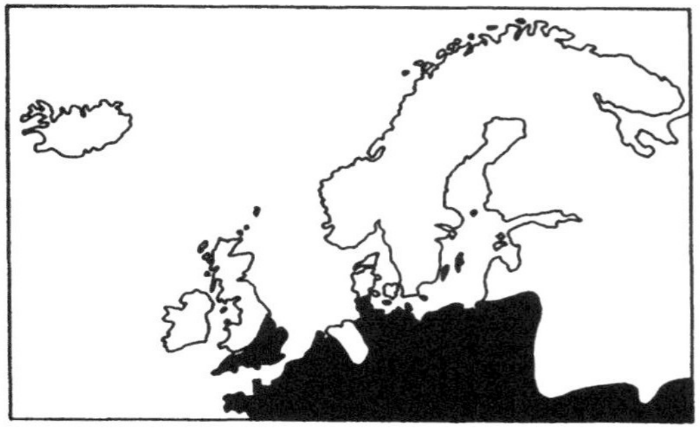

Map 32 Distribution of *Ulmus minor*

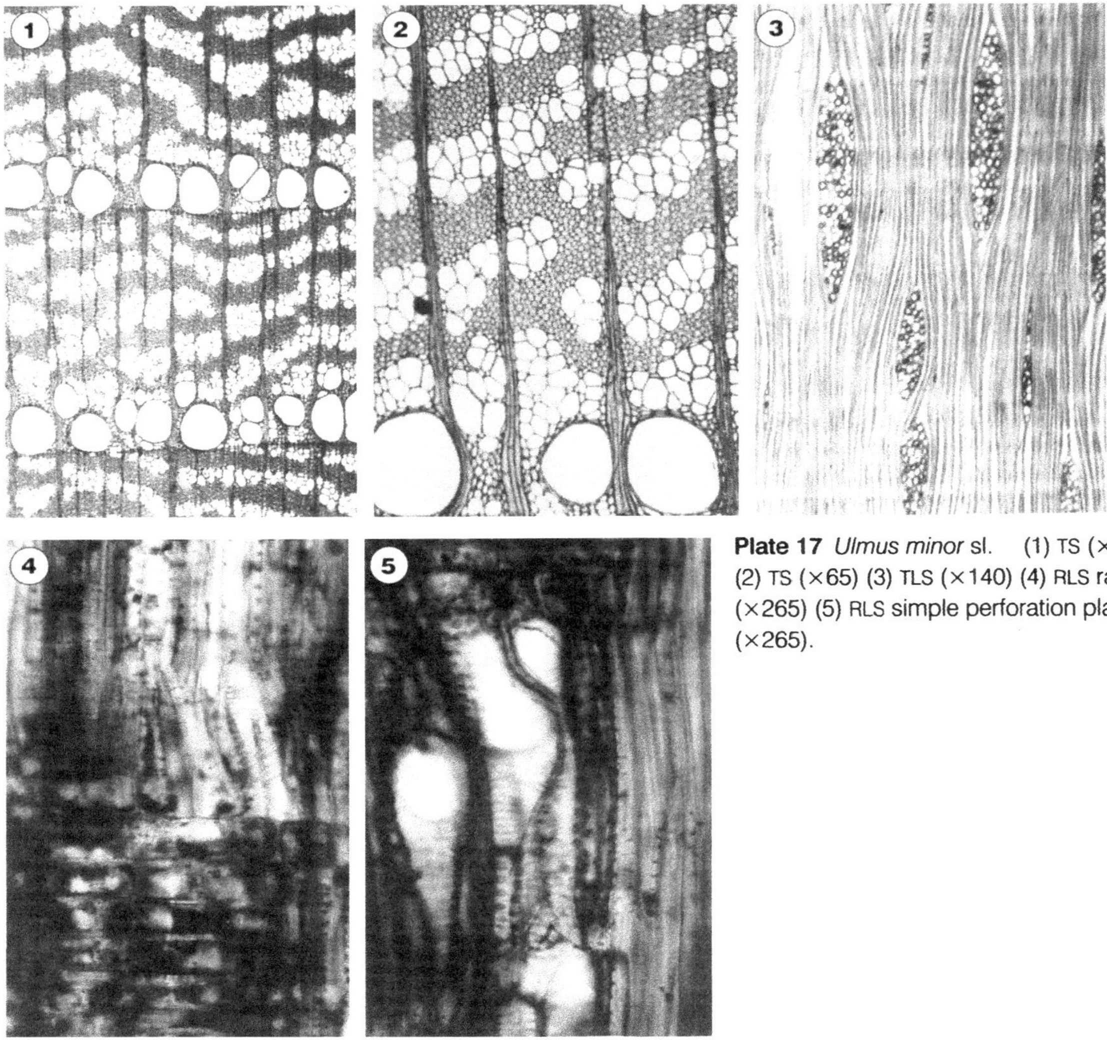

Plate 17 *Ulmus minor* sl. (1) TS (×28) (2) TS (×65) (3) TLS (×140) (4) RLS ray (×265) (5) RLS simple perforation plate (×265).

Table 2.19 Anatomical characters of *Ulmus* (Ulmaceae)

Character	*Ulmus*
TS	
Vessel arrangement	ring porous
Vessel grouping	solitary pores tangential bands
Tyloses	present
Parenchyma	paratracheal
Ground tissue	thick-walled
Ring boundary	distinct
Ray/ring boundary	rays distended
TLS	
Ray width	(1–)4–5+
Ray height	10–50
Sheath cells	absent
RLS	
Ray anatomy	homogeneous, heterogeneous type 3
Libriform fibres	present
Fibre tracheids	absent
Vascular tracheids	present
Spiral thickening	present
Perforation plates	simple

Group 4 – Diffuse porous woods with mostly solitary pores

Most of the woody plants in this group are small to medium shrubs or small trees common to woodland margins and hedgerows, such as *Cornus*, *Viburnum*, *Euonymus* and *Buxus*. The group also contains the sub-family Maloideae which includes the fruit trees, apple and pear as well as the whitebeams and hawthorns. Also within this group there is one large forest tree, the beech, of enormous ecological importance to the central European woodlands.

General anatomy

This group is defined by having mostly solitary vessels, or pores, arranged within a year's growth, in either a diffuse porous or semi-ring porous distribution. Throughout a year's growth vessels either do not change in size of frequency, or if they do they do so very gradually and decrease in size or frequency by very little.

Table 2.20 Anatomical characters of diffuse porous woods with solitary vessels

Taxon	Vessel grouping	Unisereate rays present	Largest ray width	Perforation plates	Spiral thickening present
Euonymus europaeus, E. verrucosus	solitary	yes	1	simple	yes
Caluna vulgaris	solitary	yes	1	simple (sclrfm)	no
Buxus sempervirens	solitary	yes	2	scalariform	no
Viburnum lantana	solitary	yes	3	scalariform	yes
Viburnum opulus	solitary	yes	3	scalariform	no
Cornus mas	solitary	yes	5	scalariform	no
Cornus sanguinea	solitary	yes	5	scalariform	no
Crategus, Cydonia, Pyrus, Malus, Sorbus, Mespilus	solitary	rare	3	simple	yes (fine)
Ligustrum vulgare	solitary	yes	2	simple	yes
Lonicera periclymenum	solitary (clusters)	yes	3	simple (sclfm)	yes (fine)
Lonicera xylosteum, nigra	solitary (clusters)	yes	3	simple (sclfm)	yes (fine)
Fagus sylvatica	solitary	yes	20	simple/ scalariform	no

There is one ring porous wood that has late wood anatomy that belongs to this group. This is the genus *Rosa*. It has rays that are very high though not particularly wide and so could only be confused with Fagus sylvatics since all other members of this group have very much smaller rays. Large rays in *Fagus* are much wider than those of *Rosa* and are also mostly homogeneous. *Fagus* also has both scalariform and simple perforation plates whereas *Rosa* has only simple perforation plates. Confusion is unlikely.

Taxa descriptions

Group key

Unisereate rays only	*Euonymus Calluna*
Unisereate rays and rays up to 5-sereate	*Buxus, Viburnum, Cornus* Maloideae, *Ligustrum, Lonicera*
Rays up to 20 sereate	*Fagus sylvatica*

Unisereate rays only

Euonymus europaeus (and *E. verucosus*) – Spindle tree (Celastraceae)

Plate 18

Euonymus europaeus is a small tree of hedgerows and woodland margins with a distribution across northern Europe but not in Scandinavia (see Map 33). *E. verucosus* has a more easterly distribution (see Map 34). The two species cannot be separated by their wood anatomy.

The distribution of vessels is mostly diffuse though there is a tendency towards a semi-ring porous distribution. Vessels have a consistent distribution and frequency across a year's growth, though reduce in size gradually. The annual ring is distinct and marked by several rows of flattened cells. Parenchyma is apotracheal and sparse. Rays are unisereate, homogeneous and up to 15 cells high. Rare heterogeneous rays are present. Perforation plates are simple and spiral thickening is present.

There is possible confusion with *Calluna vulgaris* (see below). Both have similar distribution of pores, distinct ring boundaries and unisereate rays. However the homogeneous rays of *Euonymus* have exclusively procumbent cells whereas those of *Calluna* have exclusively upright cells. Also scalariform perforation plates are present in *Calluna*.

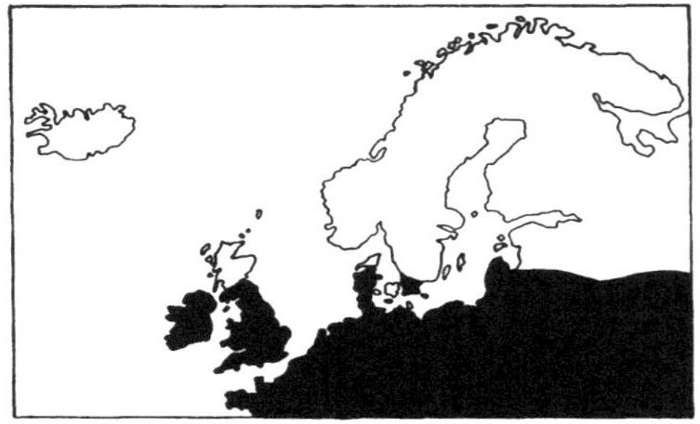

Map 33 Distribution of *Euonymus europaeus*

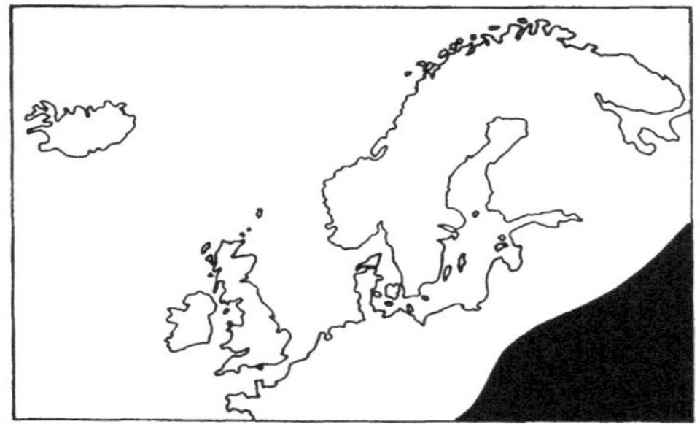

Map 34 Distribution of *Euonymus verucosus*

Table 2.21 Anatomical characters of *Euonymus europaeus* (Celastraceae)

Character	*Euonymus europaeus*	Character	*Euonymus europaeus*
TS		RLS	
Vessel arrangement	diffuse porous to semi-ring porous	Ray anatomy	homogeneous (heterogeneous type 3)
Vessel grouping	solitary	Libriform fibres	present
Tyloses	absent	Fibre tracheids	present
Parenchyma	apotracheal diffuse, sparse	Vascular tracheids	absent
Ground tissue	–	Spiral thickening	present
Ring boundary	distinct	Perforation plates	simple
Ray/ring boundary	–		
TLS			
Ray width	1		
Ray height	5–15		
Sheath cells	absent		

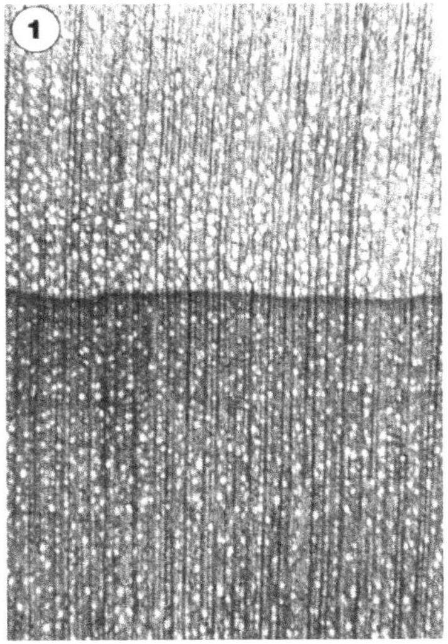
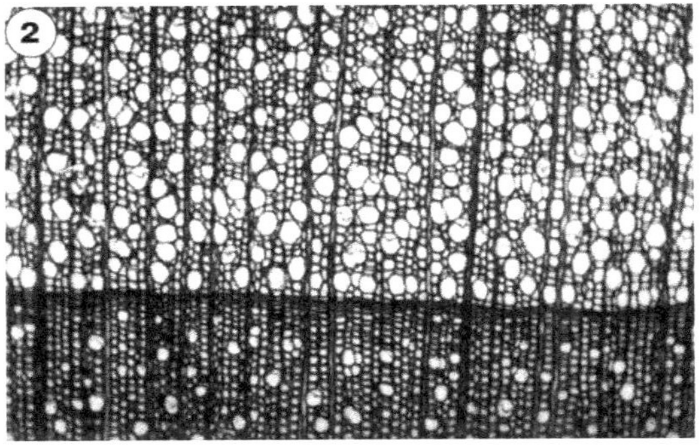

Plate 18 *Euonymus europaeus* (1) TS (×28) (2) TS (×65) (3) TLS (×65) (4) TLS (×265) (5) RLS simple perforation plate (×265) (6) RLS simple perforation plate (×265) (7) RLS simple perforation plate and ray (×265)

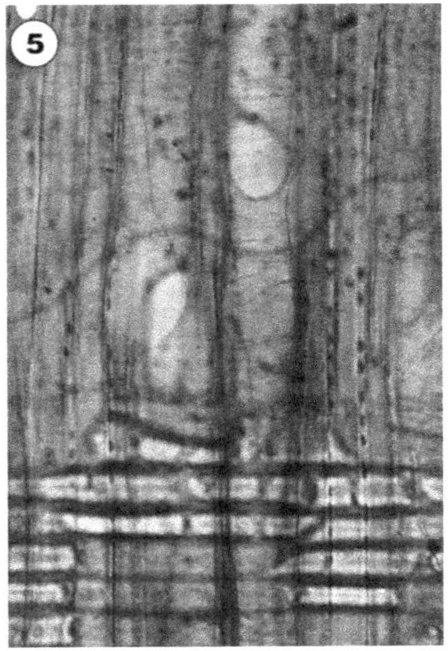
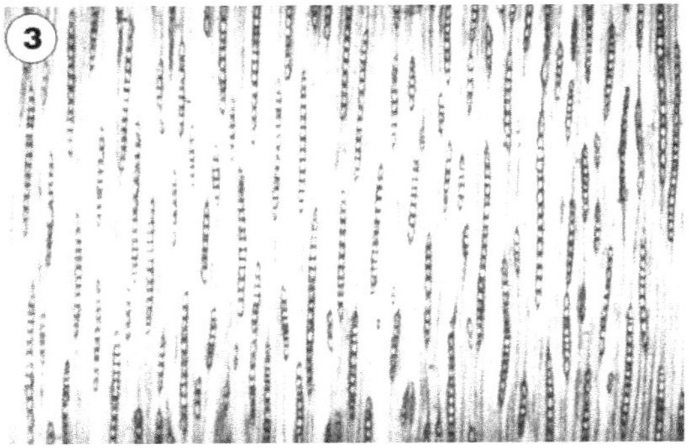
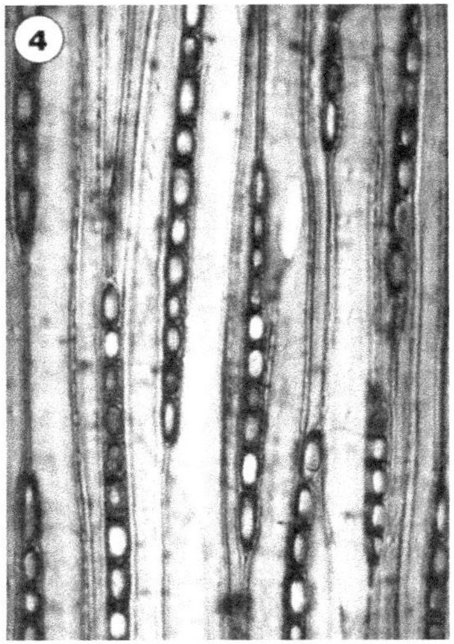
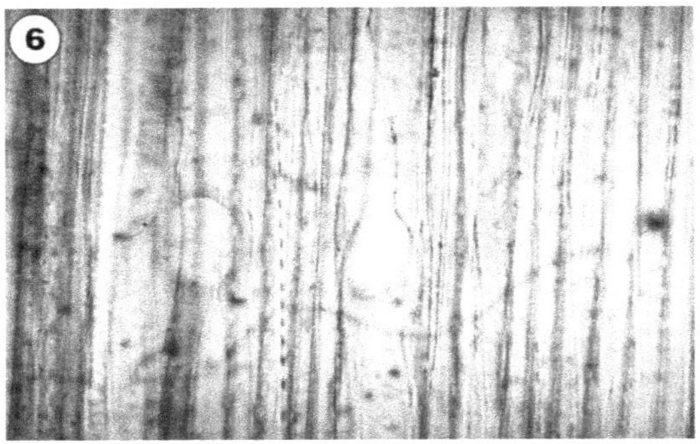
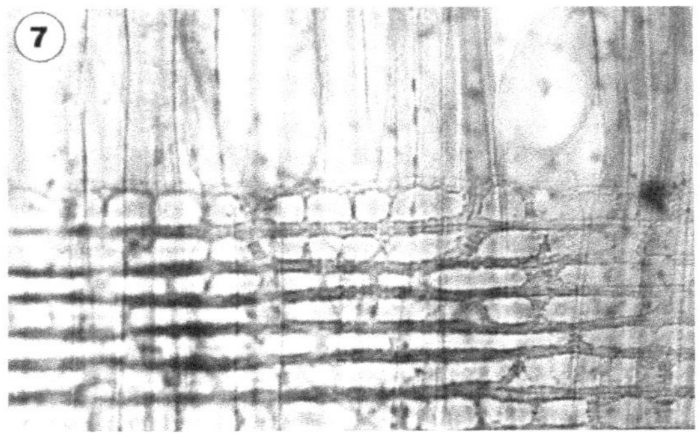

Calluna vulgaris – Heather (Ericaceae)

Plate 19

Calluna is a small to medium sized shrub with a broad distribution over most of northern Europe and Scandinavia. It is one of relatively few taxa native to Iceland (see Map 35).

Calluna has an entirely diffuse arrangement of pores with those of the largest diameter appearing about two-thirds to three-quarters across a year's growth. The ring boundary is made distinct by thick-walled fibres in the late wood. Rays are uniserate, homogeneous and made up of upright or fusiform cells. Perforation plates are mostly simple though scalariform plates with one to five bars are present. There is no spiral thickening.

There is possible confusion with *Euonymous*, though ray anatomy is different and spiral thickening is absent (see also *Euonymus* above).

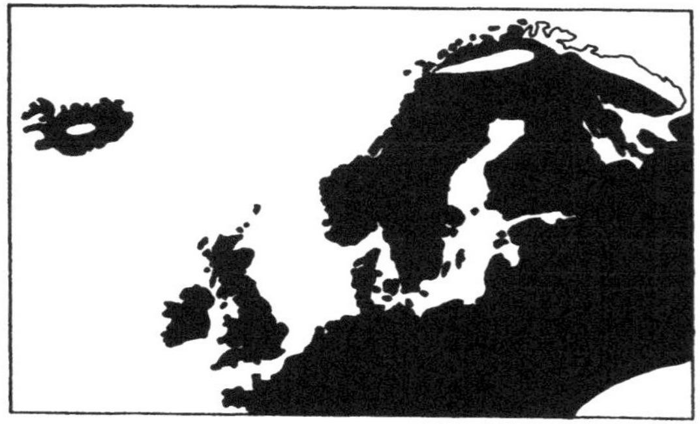

Map 35 Distribution of *Calluna vulgaris*

Table 2.22 Anatomical characters of *Calluna vulgaris* (Ericaceae)

Character	*Calluna vulgaris*	Character	*Calluna vulgaris*
TS		RLS	
Vessel arrangement	diffuse porous	Ray anatomy	homogeneous
Vessel grouping	solitary	Ray / vessel pits	small, round
Tyloses	absent	Libriform fibres	absent
Parenchyma	apotracheal diffuse	Fibre tracheids	present
Ground tissue	–	Vascular tracheids	absent
Ring boundary	distinct	Spiral thickening	absent
Ray/ring boundary	–	Crystals	absent
TLS		Perforation plates	simple (scalariform, 1–5)
Ray width	1		
Ray height	1–5		
Ray cell shape	upright – fusiform		
Sheath cells	absent		

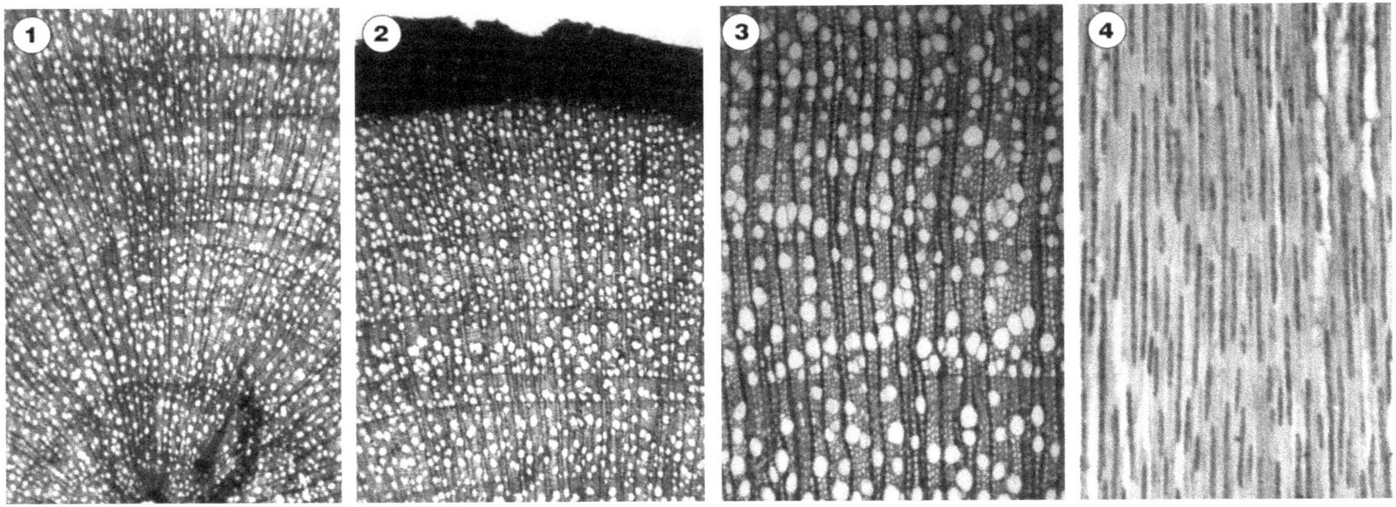

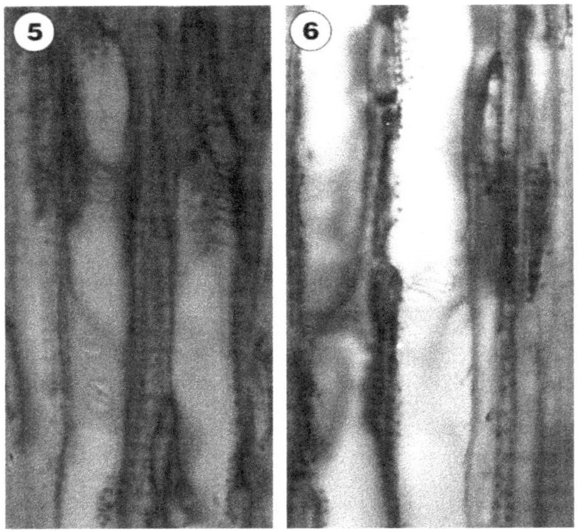

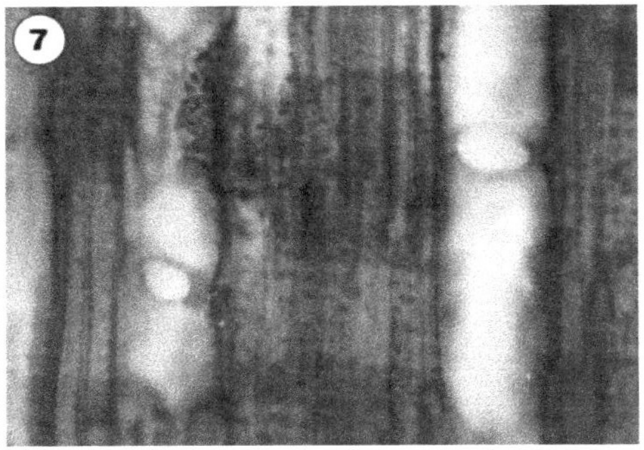

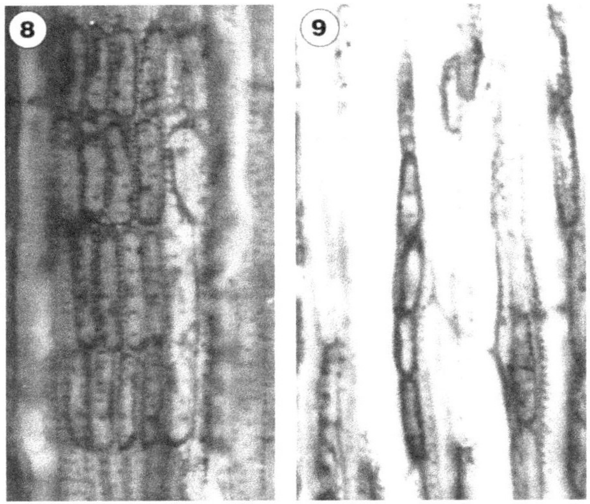

Plate 19 *Calluna vulgaris* (1) TS (×28) (2) TS (×28) (3) TS (×65) (4) TLS (×65) (5) RLS scalariform perforation plate (×265) (6) RLS scalariform perforation plate (×265) (7) RLS simple perforation plates (×265) (8) RLS (×265) (9) RLS (×265).

Unisereate rays present and rays up to 5 sereate

Buxus sempervirens – Box (Buxaceae)

Plate 20

Buxus is a large shrub or small tree native to only a few small and isolated areas of western northern Europe, but with a broader distribution further south and in the northern Mediterranean area (see Map 36). It is an important wood for small carved and turned artefacts due to its very dense clear/creamy white coloured wood. There is archaeological evidence for both long distance trade in the wood and for its cultivation outside of its native distribution.

Buxus has a very dense appearance with thick-walled fibres and very small vessels. The vessels are all solitary and with a diffuse porous arrangement. There is little difference in vessel diameter across a year's growth and, as such, the ring boundary tends to be a little indistinct. Parenchyma is apotracheal and diffuse. Rays are heterogeneous mostly bisereate, occasionally unisereate and up to 12 cells high. Upright cells on the ray margins are obvious in TLS. Perforation plates are scalariform and often small with up to 12 bars.

The very small size of the pores, the thick-walled ground tissue and small scalariform perforation plates make confusion of this species with others unlikely, however see *Viburnum*, *Cornus*, Maloideae, *Ligustrum*, *Lonicera* below.

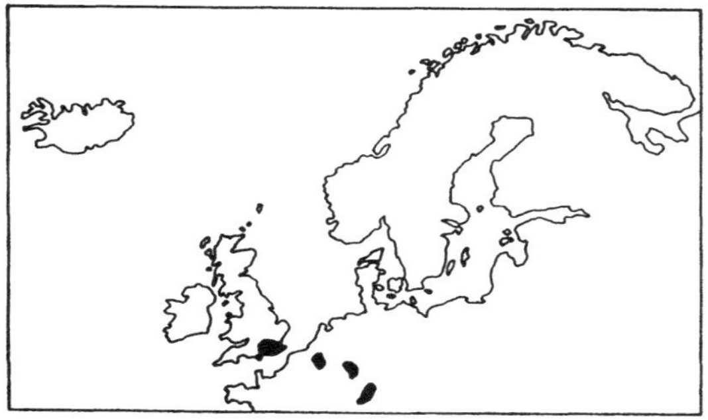

Map 36 Distribution of *Buxus sempervirens*

Table 2.23 Anatomical characters of *Buxus sempervirens* (Buxaceae)

Character	*Buxus sempervirens*	Character	*Buxus sempervirens*
TS		RLS	
Vessel arrangement	diffuse porous	Ray anatomy	heterogeneous 3 (1, 2)
Vessel grouping	solitary	Libriform fibres	absent
Tyloses	absent	Fibre tracheids	present
Parenchyma	apotracheal diffuse	Vascular tracheids	absent
Ground tissue	thick-walled	Spiral thickening	absent
Ring boundary	indistinct	Perforation plates	scalariform, 5–12
Ray/ring boundary	–		
TLS			
Ray width	(1–)2		
Ray height	6–12		
Sheath cells	absent		

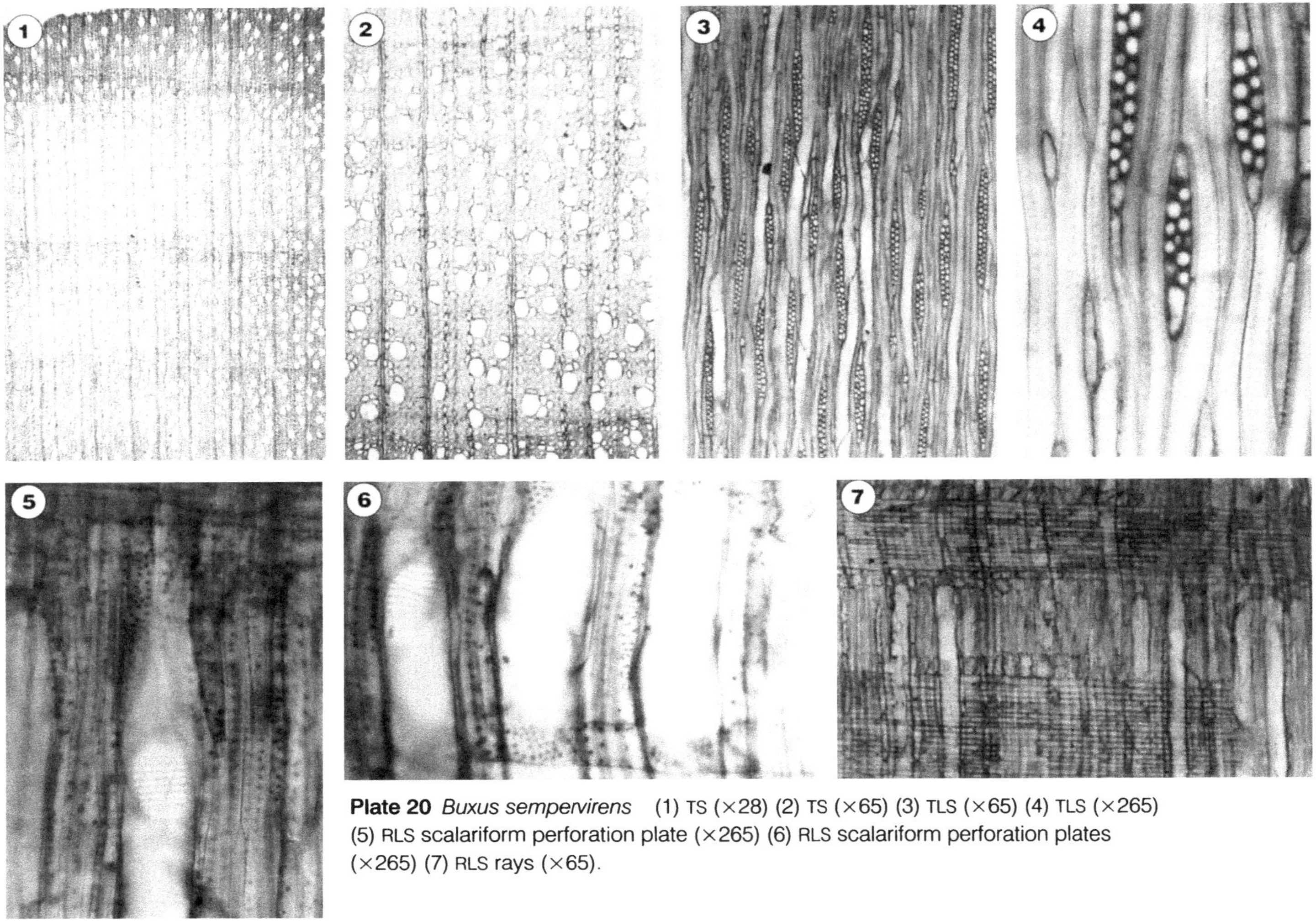

Plate 20 *Buxus sempervirens* (1) TS (×28) (2) TS (×65) (3) TLS (×65) (4) TLS (×265) (5) RLS scalariform perforation plate (×265) (6) RLS scalariform perforation plates (×265) (7) RLS rays (×65).

Viburnum opulus – Guelder rose

Viburnum lantana – Wayfaring tree (Caprifoliaceae)

Plate 21 and Plate 22

Viburnum opulus and *V. lantana* are both large shrubs or small trees of hedgerows and woodland margins. *Viburnum opulus* has a distribution across northern Europe and into Scandinavia (see Map 37) while *V. lantana* is only native to western France and southern Britain in this region and is more broadly distributed in southern Europe (see Map 38).

Both taxa have a diffuse porous arrangement of vessels with a slight decrease in vessel diameter across a year's growth. Parenchyma is apotracheal. Rays are heterogeneous, with many rows of upright cells, up to 3 cells wide and 15 cells high. Unisereate rays are made up of upright cells only. Sheath cells are present in larger rays. Perforation plates are tall, scalariform and often with over 25 bars. Spiral thickening is present in *Viburnum lantana* only and this is the only character that allows these two species to be separated.

There is possible confusion with other taxa with a diffuse porous arrangement of vessels and small multisereate rays, though Malaloideae, *Ligustrum* and *Lonicera* have simple perforation plates and *Buxus* has rather smaller rays and much shorter perforation plates. *Cornus* has very similar perforation plates but can be separated from *Viburnum* by its much higher rays.

Map 37 Distribution of *Viburnum opulus*

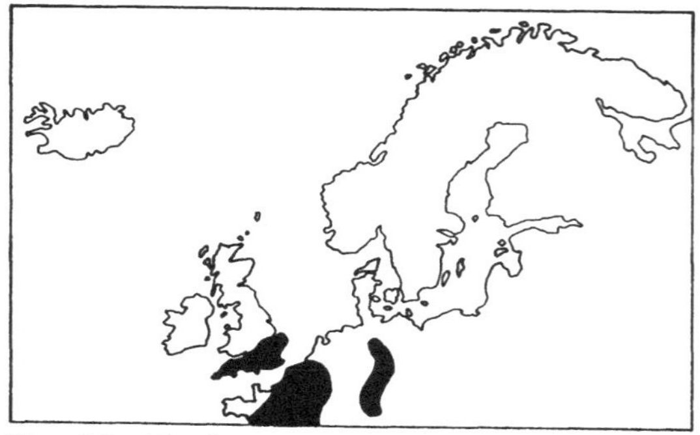

Map 38 Distribution of *Viburnum lantana*

Table 2.24 Anatomical characters of *Viburnum opulus* and *V. lantana* (Caprifoliaceae)

Character	*Viburnum opulus*	*Viburnum lantana*
TS		
Vessel arrangement	diffuse porous	diffuse porous
Vessel grouping	solitary	solitary
Tyloses	absent	absent
Parenchyma	apotracheal diffuse	apotracheal diffuse
Ground tissue	–	–
Ring boundary	± indistinct	± indistinct
Ray/ring boundary	–	–
TLS		
Ray width	1–3	1–3
Ray height	5–15	5–15
Sheath cells	present	present
RLS		
Ray anatomy	heterogeneous type 1	heterogeneous type 1
Libriform fibres	absent	absent
Fibre tracheids	present	present
Vascular tracheids	absent	absent
Spiral thickening	absent	present
Perforation plates	scalariform ±25 +	scalariform ±25 +

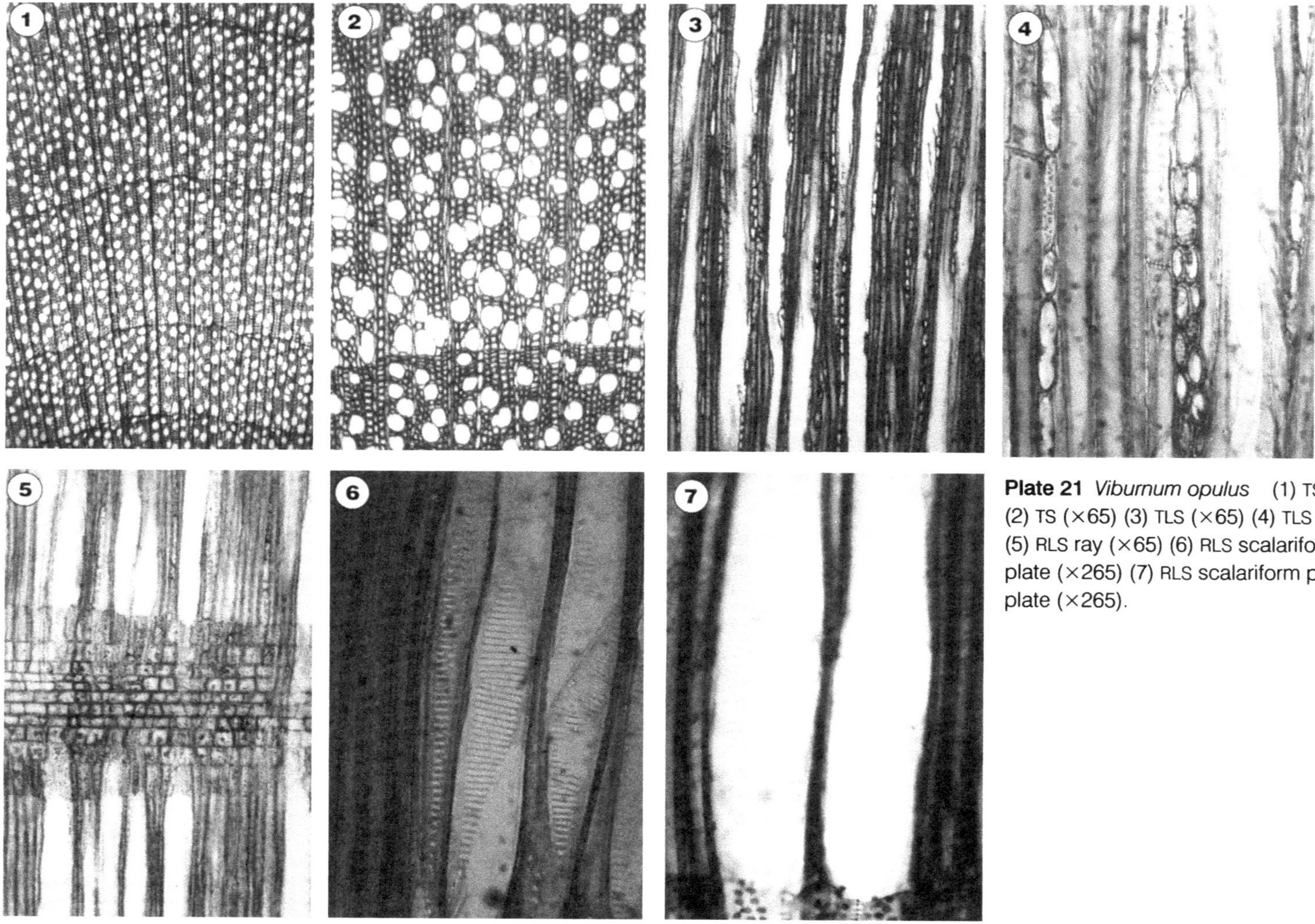

Plate 21 *Viburnum opulus* (1) TS (×28) (2) TS (×65) (3) TLS (×65) (4) TLS (×265) (5) RLS ray (×65) (6) RLS scalariform perforation plate (×265) (7) RLS scalariform perforation plate (×265).

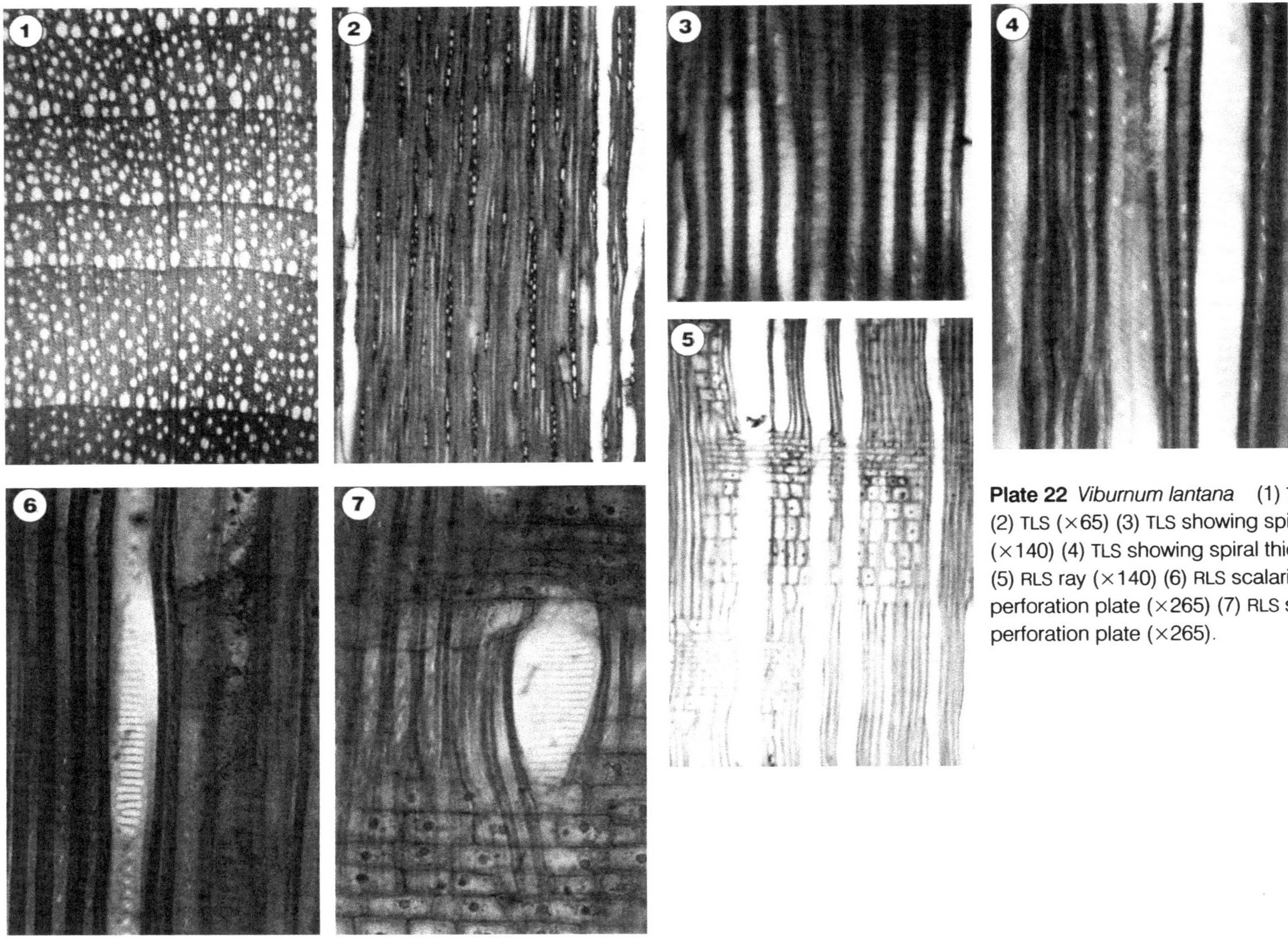

Plate 22 *Viburnum lantana* (1) TS (×28) (2) TLS (×65) (3) TLS showing spiral thickening (×140) (4) TLS showing spiral thickening (×265) (5) RLS ray (×140) (6) RLS scalariform perforation plate (×265) (7) RLS scalariform perforation plate (×265).

Cornus sanguinea – Dogwood

Cornus mas – Cornelian cherry (Cornaceae)

Plate 23 and Plate 24

Both *Cornus* species are shrubs or small trees of hedgerows and woodland margins. Both have a broad northern European distribution though *Cornus sanguinea* reaches further to the north and into Scandinavia (see Maps 39 and 40).

Cornus has a diffuse porous distribution of vessels with a more or less constant vessel diameter across a year's growth. Ring boundaries are relatively distinct and gently undulating. Parenchyma is apotrachael diffuse and in more or less tangential bands. Ground tissue is particularly thick-walled. Rays are heterogeneous with many rows of square and upright cells, up to 7 cells wide though often no more than 5, and up to 50 cells high. Unisereate rays are composed entirely of upright cells. Perforation plates are tall, scalariform and have up to 30 bars, though often fewer.

Schweingruber (1990) notes that the two *Cornus* species may be separated on the basis of the growth ring boundaries, which are less distinct in *Cornus mas* than in *C. sanguinea*, pore density which is greater in *Cornus sanguinea*, and the ground tissue which is more distinctly thicker-walled in *Cornus mas*. Differentiation is probably difficult in archaeological material.

There is possible confusion with other diffuse porous woods, but especially *Viburnum* which has similar characters, however *Cornus* has much higher rays.

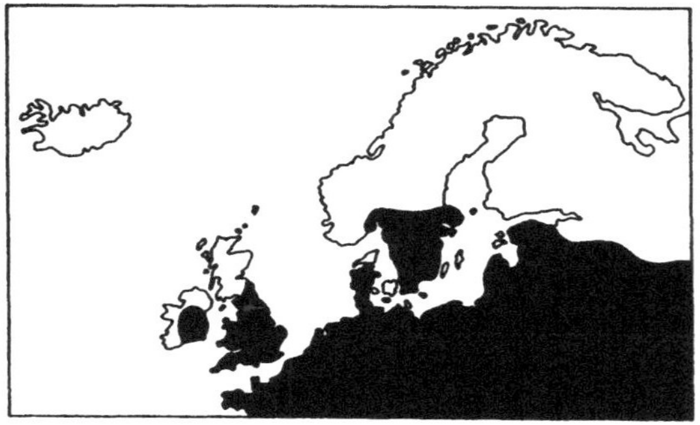

Map 39 Distribution of *Cornus sanguinea*

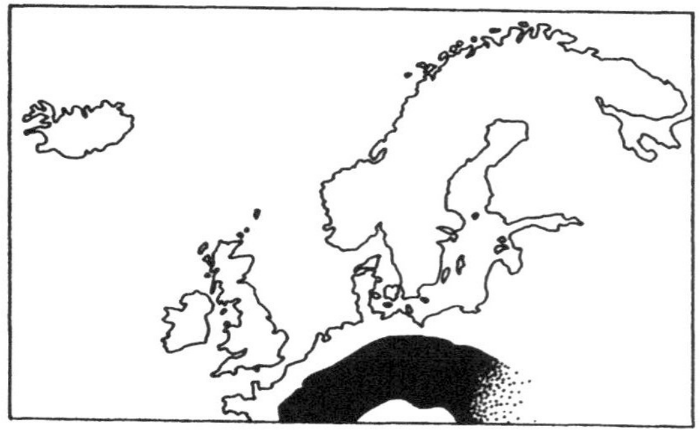

Map 40 Distribution of *Cornus mas*

Table 2.25 Anatomical characters of *Cornus* (Cornaceae)

Character	*Cornus sanguinea*	*Cornus mas*
TS		
Vessel arrangement	diffuse porous	diffuse porous
Vessel grouping	solitary – dense	solitary – sparse
Tyloses	absent	absent
Parenchyma	apotracheal diffuse and ± tangential bands (paratracheal)	apotracheal diffuse and ± tangential bands (paratracheal)
Ground tissue	± thick-walled	distinctly thick-walled
Ring boundary	distinct and undulating	± indistinct and undulating
Ray/ring boundary	–	–
TLS		
Ray width	1, 3–5(–7)	1, 3–5(–7)
Ray height	5–50	5–50
Sheath cells	(present)	(present)
RLS		
Ray anatomy	heterogeneous type 1	heterogeneous type 1
Libriform fibres	absent	absent
Fibre tracheids	present	present
Vascular tracheids	absent	absent
Spiral thickening	absent	absent
Perforation plates	scalariform ±20+	scalariform ±20+

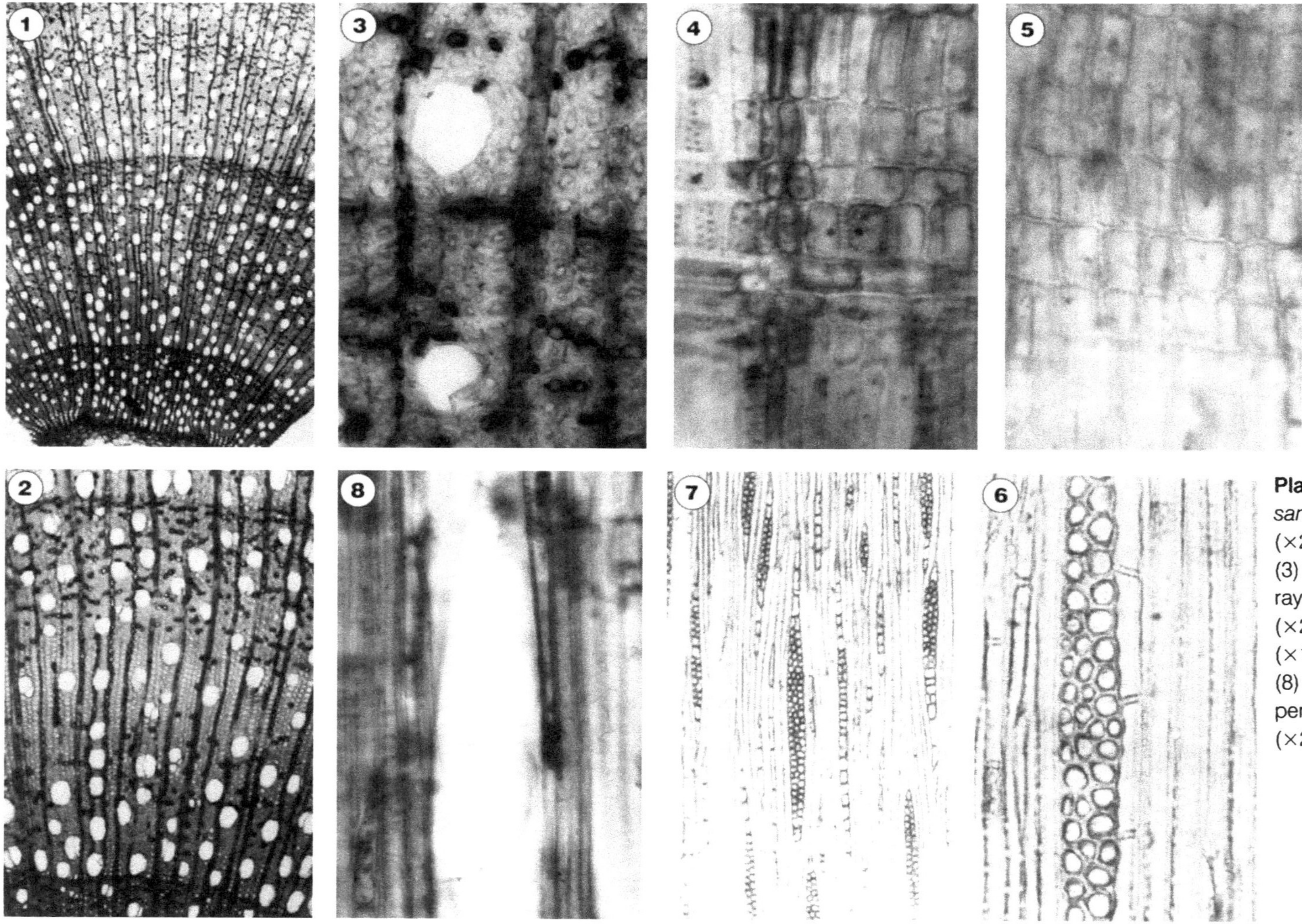

Plate 23 *Cornus sanguinea* (1) TS (×25) (2) TS (×65) (3) TS (×265) (4) RLS ray (×265) (5) RLS ray (×265) (6) TLS (×140) (7) TLS (×65) (8) RLS scalariform perforation plate (×265).

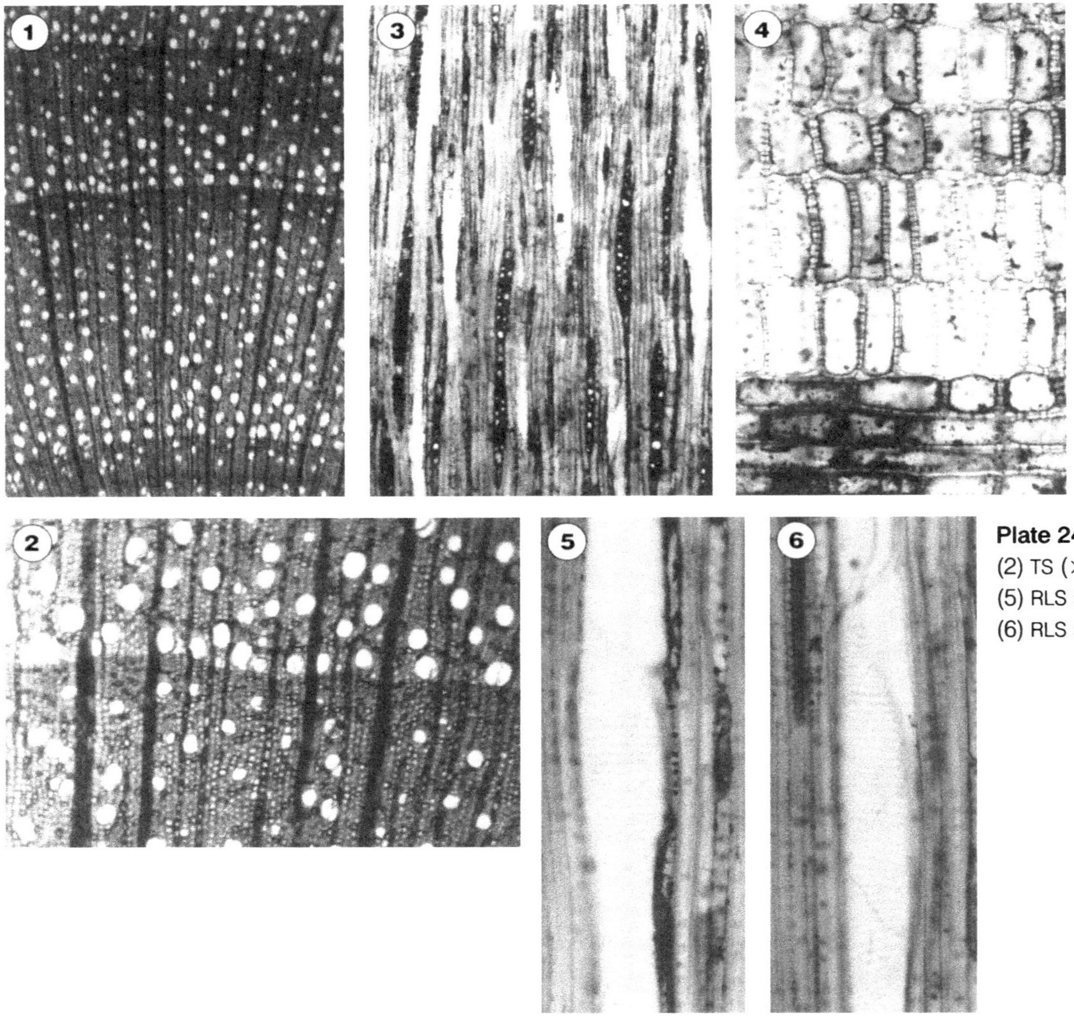

Plate 24 *Cornus mas* (1) TS (×28) (2) TS (×65) (3) TLS (×65) (4) RLS ray (×265) (5) RLS scalariform perforation plate (×265) (6) RLS scalariform perforation plate (×265).

Maloideae (Rosaceae)

Includes: *Malus* – apple *Sorbus* – whitebeams
Pyrus – pear *Crategus* – hawthorns

Plate 25 and Plate 26

The Maloideae is a sub-family group within the Rosaceae and includes some important and widespread trees, often used for their fruit, though their wood is also used in artefact manufacture. The distributions of the following have been included:

Crategus calycina (Map 41)
Crategus monogyna – hawthorn (Map 42)
Crategus laevigata – Midland hawthorn (Map 43)
Amelanchoir ovalis – Juneberry (Map 44)
Pyrus pyraster – wild pear (Map 45)
Pyrus cordata – Plymouth pear (Map 46)
Malus sylvestris – apple (Map 47)
Sorbus aria – common whitebeam (Map 48)
Sorbus rupicola (Map 49)
Sorbus aucuparia – rowan (Map 50)
Sorbus torminalis – wild service tree (Map 51)
Sorbus hybrida – Swedish service tree (Map 52)
Sorbus intemedia – Swedish whitebeam (Map 53)

Sorbus is a very complex genus with many apomictic in addition to well-defined sexual species. There are numerous species that have arisen apomicticly and have very limited and often isolated distributions in both Britain and in Scandinavia.

Maloideae as a group displays very little variation in its anatomy. It is diffuse porous though often vessels in the early wood are larger and more densely packed than those in the late wood. Parenchyma is apotracheal and the ground tissue is most often thick-walled. Gummy deposits occur in some vessels. Rays are 2 to 3 cells wide and up to 15 cells high, mostly homogeneous and rather uniform in both size and shape. Unisereate rays are absent or rare. Perforation plates are simple though scalariform and reticulate

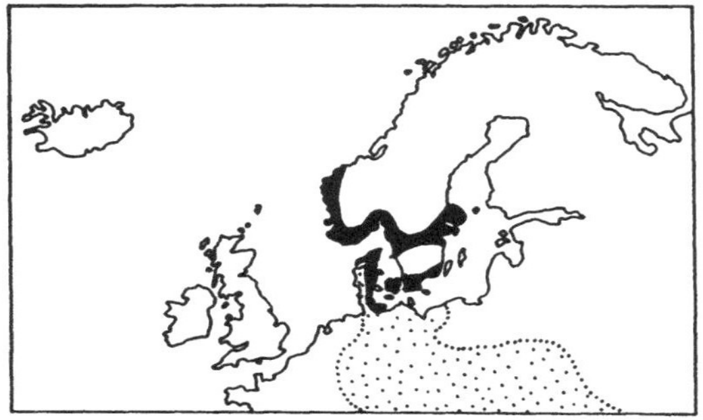

Map 41 Distribution of *Crategus calycina*

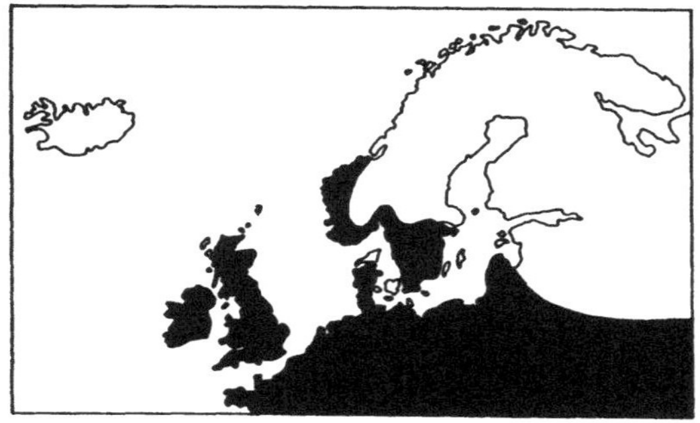

Map 42 Distribution of *Crategus monogyna*

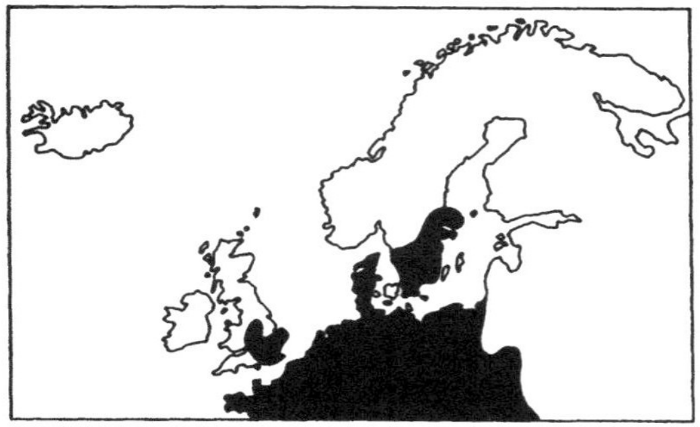

Map 43 Distribution of *Crategus laevigata*

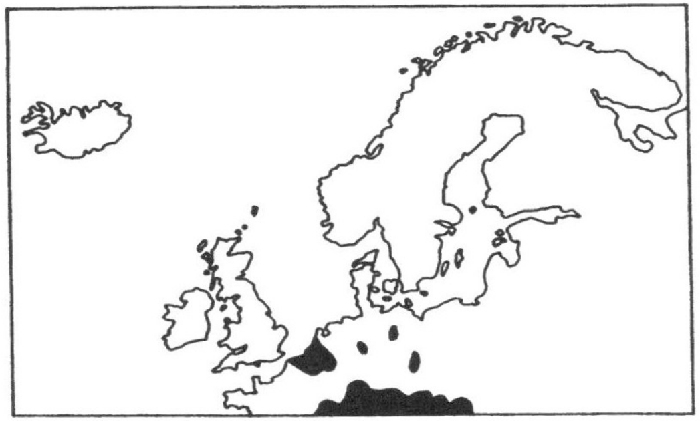

Map 44 Distribution of *Amelanchoir ovalis*

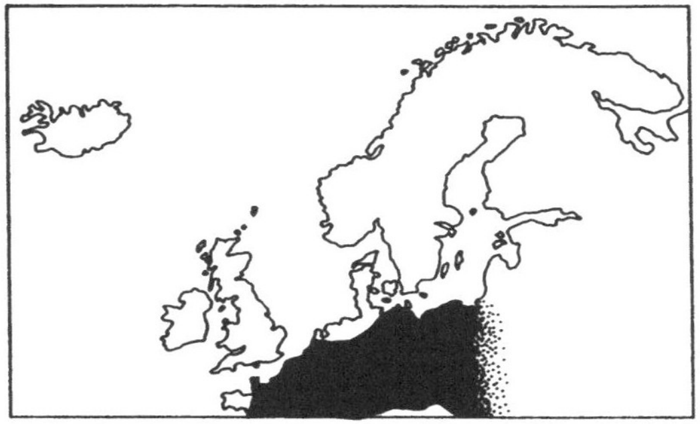

Map 45 Distribution of *Pyrus pyraster*

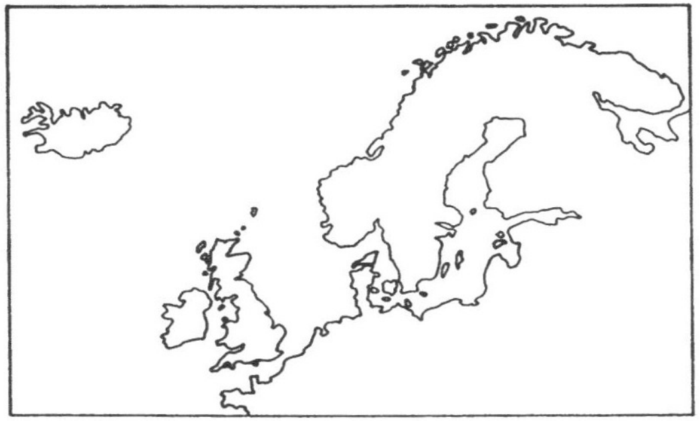

Map 46 Distribution of *Pyrus cordata*

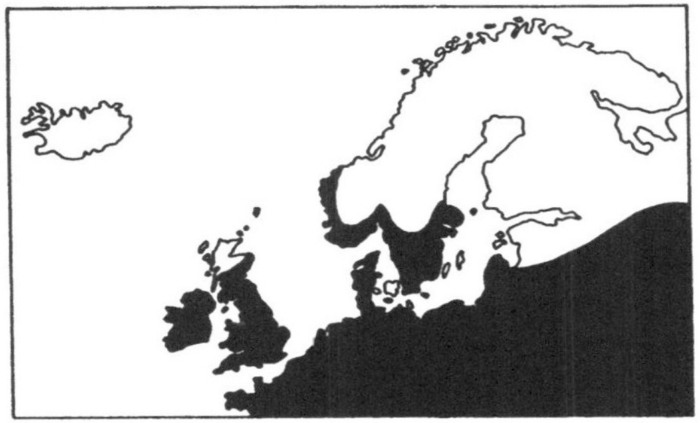

Map 47 Distribution of *Malus sylvestris*

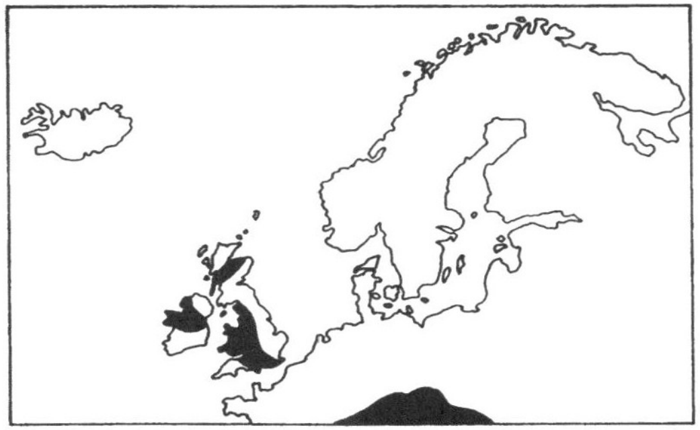

Map 48 Distribution of *Sorbus aria*

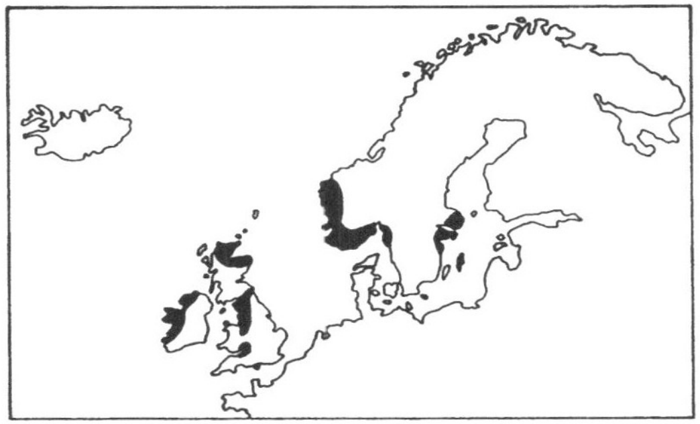

Map 49 Distribution of *Sorbus rupicola*

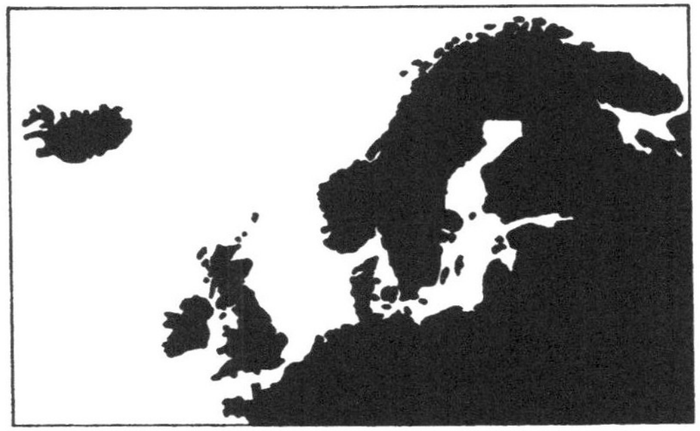

Map 50 Distribution of *Sorbus aucuparia*

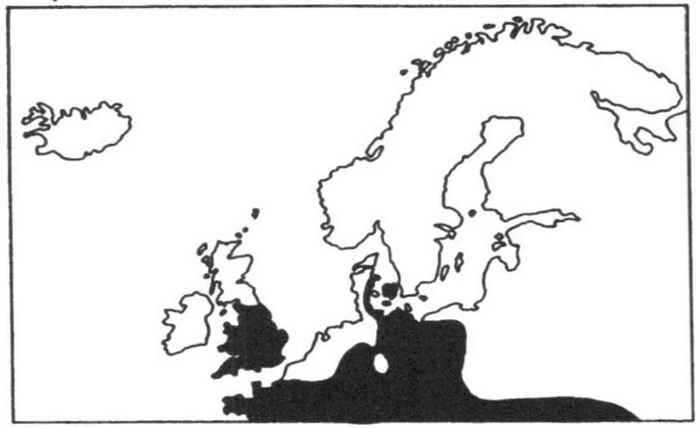

Map 51 Distribution of *Sorbus torminalis*

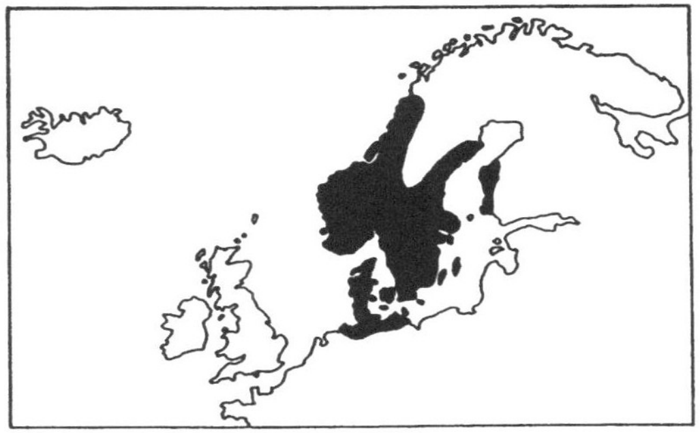

Map 52 Distribution of *Sorbus hybrida*

perforation plates are very occasionally present. Spiral thickening is present though fine.

There is possible confusion with other diffuse porous woods that have solitary vessels and small multisereate rays. The simple perforation plates and mostly multisereate homogeneous rays should differentiate this group from others.

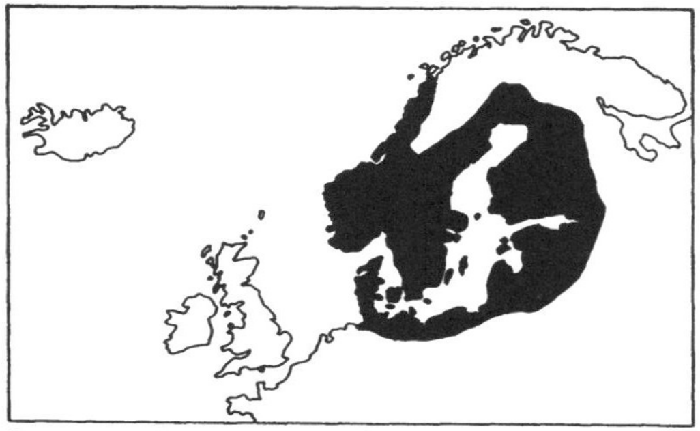

Map 53 Distribution of *Sorbus intemedia*

Table 2.26 Anatomical characters of the Maloideae (Rosaceae)1 = includes *Crategus, Sorbus, Malus, Pyrus, Amelanchoir*

Character	Maloideae
TS	
Vessel arrangement	diffuse porous (-semi-ring porous)
Vessel grouping	solitary
Tyloses	present
Parenchyma	apotracheal diffuse
Ground tissue	–
Ring boundary	mostly distinct
Ray/ring boundary	–
TLS	
Ray width	2–3
Ray height	5–15
Sheath cells	absent
RLS	
Ray anatomy	homogeneous (heterogeneous type 3)
Libriform fibres	absent
Fibre tracheids	present
Vascular tracheids	absent
Spiral thickening	(fine)
Crystals	(present)
Perforation plates	simple

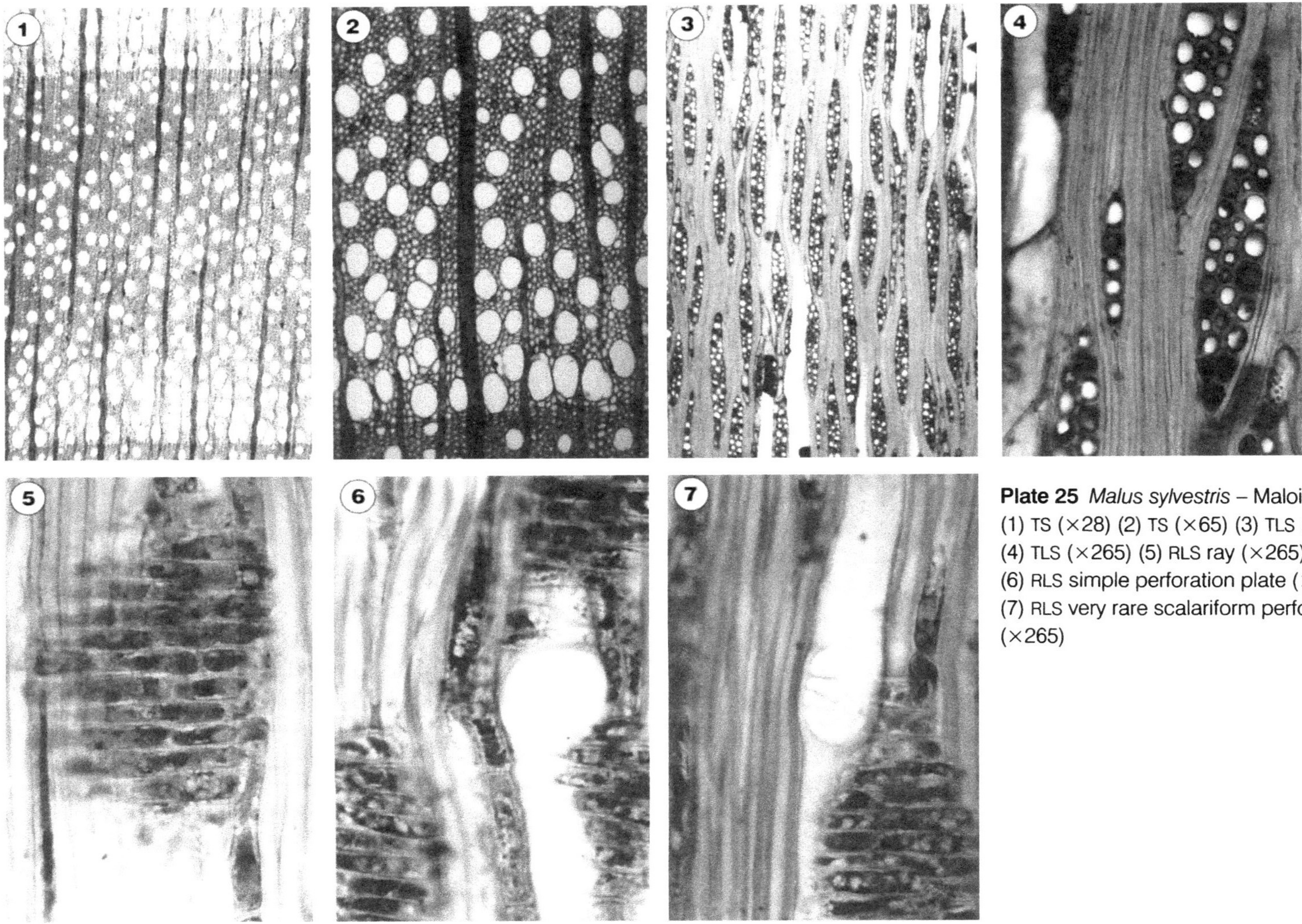

Plate 25 *Malus sylvestris* – Maloideae
(1) TS (×28) (2) TS (×65) (3) TLS (×65)
(4) TLS (×265) (5) RLS ray (×265)
(6) RLS simple perforation plate (×265)
(7) RLS very rare scalariform perforation plate (×265)

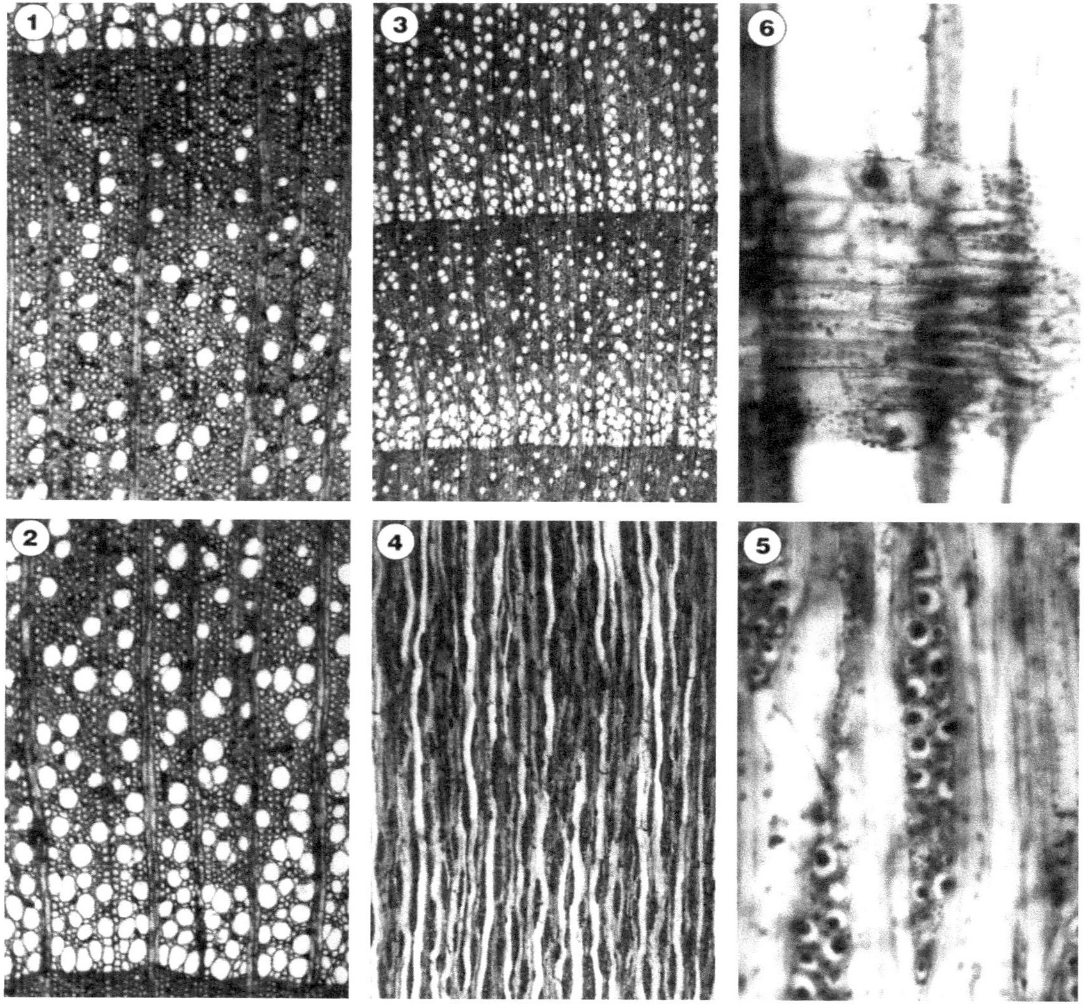

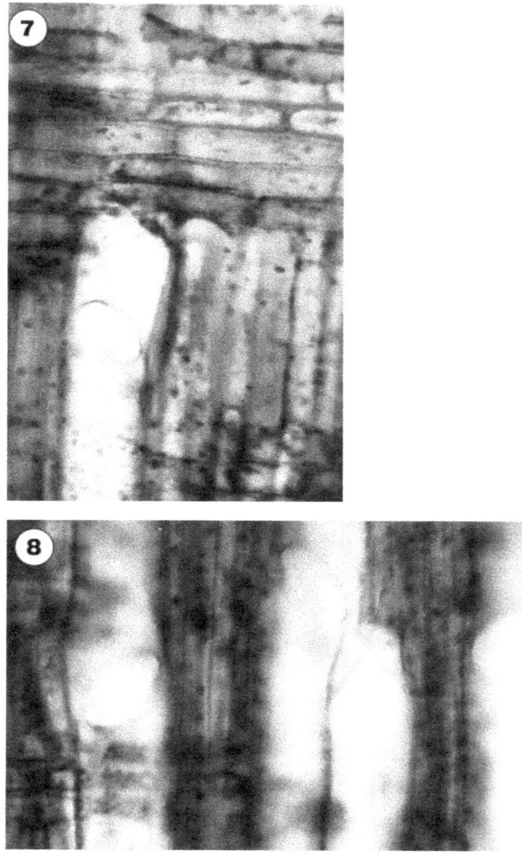

Plate 26 *Sorbus aria* – Maloideae (1) TS (×65) (2) TS (×65) (3) TS (×28) (4) TLS (×65) (5) TLS (×265) (6) RLS ray (×265) (7) RLS simple perforation plate (×265) (8) RLS simple perforation plate (×265).

Ligustrum vulgare – Privet (Oleaceae)

Plate 27

This is a small shrub of base rich woodland, which has a distribution across northern Europe and into southern Scandinavia (see Map 54).

Ligustrum has a diffuse porous distribution of vessels with a tendency towards the semi-ring porous state which makes ring boundaries rather distinct. Ground tissue is thick-walled and parenchyma is apotracheal. Rays are unisereate and bisereate, rarely larger, up to 20 cells high and heterogeneous with several rows of upright cells that are much larger than the body cells of the ray when seen in TLS. Perforation plates are simple and spiral thickening is present.

There may be confusion with Maloideae though the presence of unisereate and strongly heterogeneous rays should prevent this. Confusion with *Lonicera xylosteum*/ *L.nigra* is more likely (see below).

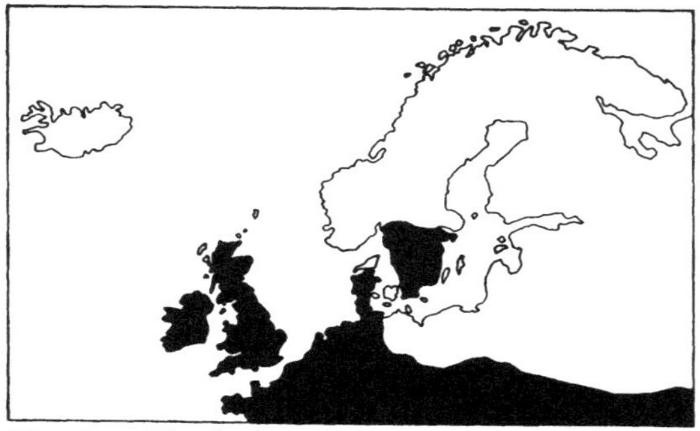

Map 54 Distribution of *Ligustrum vulgare*

Table 2.27 Anatomical characters of *Ligustrum vulgare* (Oleaceae)

Character	*Ligustrum vulgare*	Character	*Ligustrum vulgare*
TS		RLS	
Vessel arrangement	diffuse porous to semi-ring porous	Ray anatomy	heterogeneous types 1, 2, 3
Vessel grouping	solitary	Libriform fibres	present
Tyloses	absent	Fibre tracheids	present
Parenchyma	apotracheal diffuse	Vascular tracheids	absent
Ground tissue	thick-walled	Spiral thickening	present
Ring boundary	distinct	Crystals	present
Ray/ring boundary	–	Perforation plates	simple
TLS			
Ray width	1–2(–3)		
Ray height	10–20		
Sheath cells	absent		

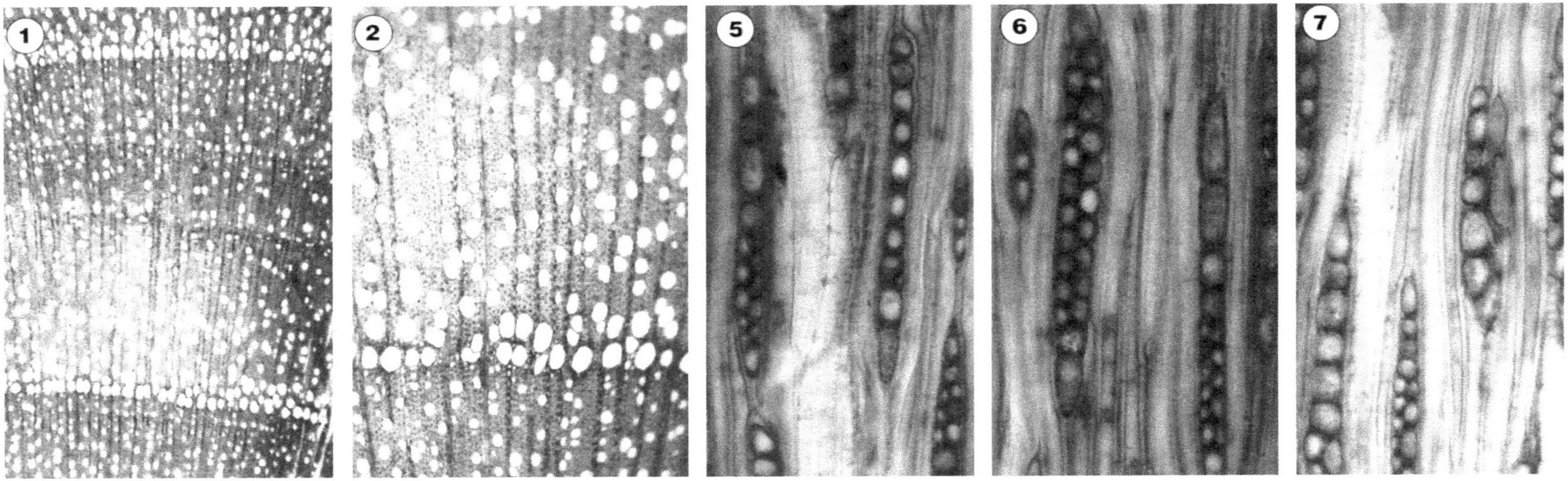
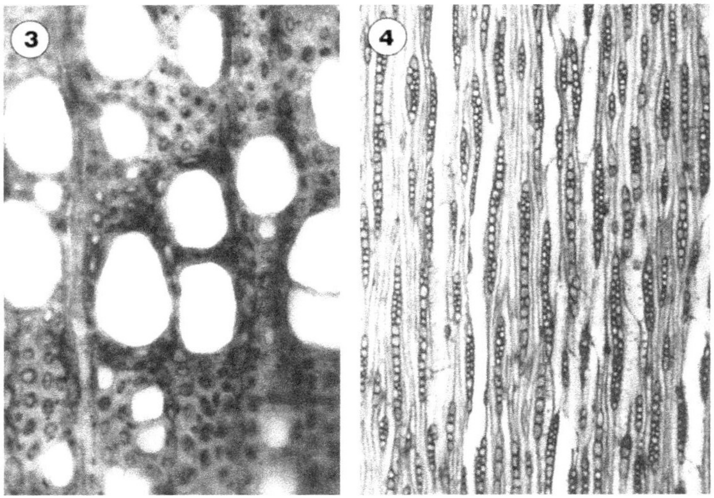
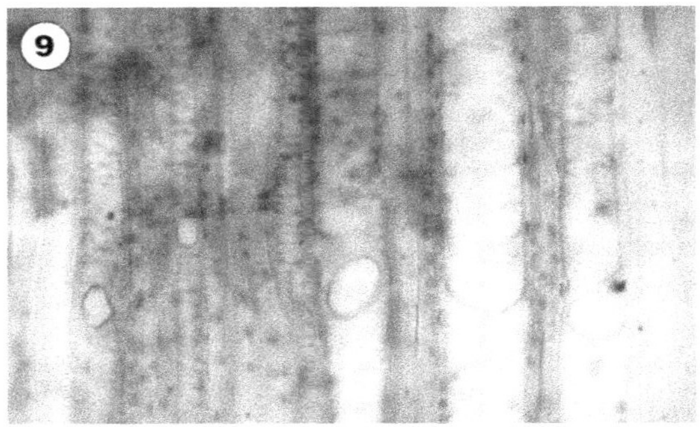

Plate 27 *Ligustrum vulgare* (1) TS (×28) (2) TS (×65) (3) TS (×265) (4) TLS (×65) (5) TLS (×265) (6) TLS (×265) (7) TLS (×265) (8) RLS ray (×265) (9) RLS simple perforation plates (×140).

Lonicera periclymenum

Lonicera xylosteum, L. nigra – Honeysuckles (Caprifoliaceae)

Plate 28 and Plate 29

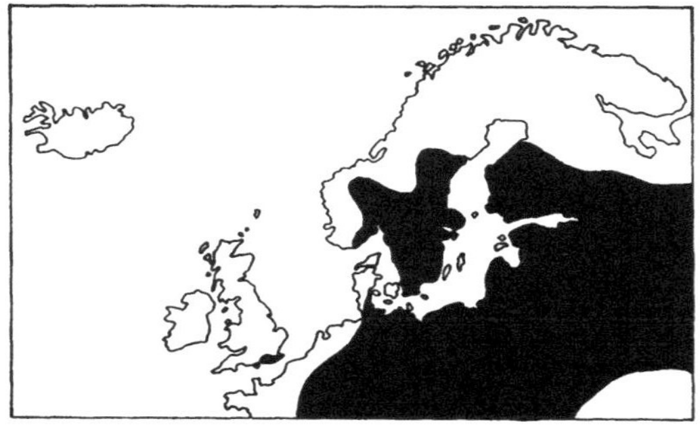

Map 55 Distribution of *Lonicera xylosteum*

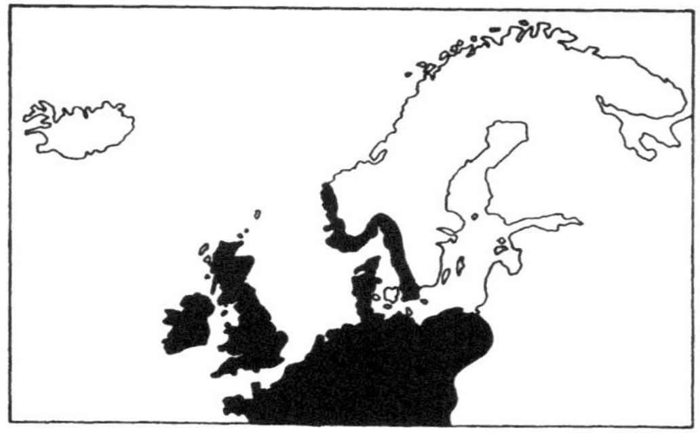

Map 56 Distribution of *Lonicera periclymenum*

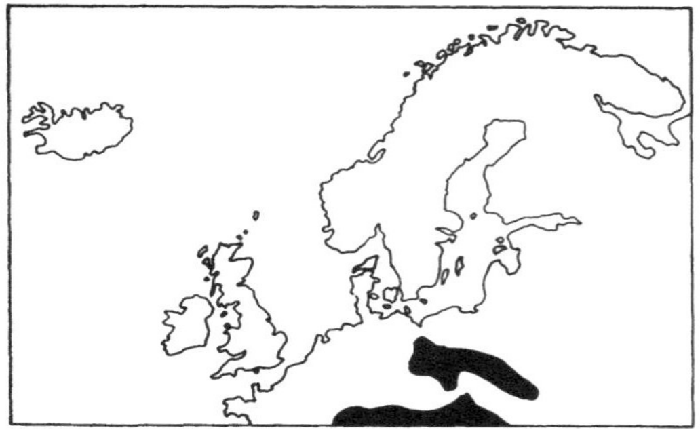

Map 57 Distribution of *Lonicera nigra*

Lonicera periclymenum is a scrambler of woodlands and hedgerows whereas both *L. xylosteum* and *L. nigra* are small shrubs of woodland margins. *L. xylosteum* has a wide ranging distribution in northern Europe from the east to central Europe and into Scandinavia, with an outlyer distribution in southern Britain (see Map 55). The other two species, *L. nigra* and *L. periclymenum* have a more limited distribution (see Maps 56 and 57).

All three species are diffuse porous although *Lonicera xylosteum* and *L. nigra* have a tendency towards the semi-ring porous state. In all three taxa pores are mostly solitary although some pore clusters occur. Rays are unisereate to 4 cells wide though most are 1 to 3 cells wide, and are strongly heterogeneous. Unisereate rays have upright cells only. In larger rays sheath cells may be present. Perforation plates are mostly simple although scalariform perforation plates may occur.

Lonicera periclymenum may be differentiated from *L. xylosteum* and *L. nigra* by the ground tissue which is thicker walled in the late wood in *L xylosteum* and *L. nigra* than in *L. periclymenum* which has more or less thin-walled ground tissue throughout. This character may be difficult to see, though where the semi-ring porous state occurs in *L.xylosteum* and *L.nigra* this is a clearer character allowing the two groups to be separated.

There may be confusion between *L. xylosteum/L. nigra* and *Ligustrum* (above) since both may be semi-ring porous and have other characters in common. The two can be separated by ray structure.

Ligustrum rays are commonly unisereate and bisereate, rarely larger. Unisereate rays are made of procumbent cells and upright marginal cells. Rays commonly have as few as one row of upright marginal cells.

Lonicera xylosteum/L. nigra has rays which are up to 4 cells wide and are commonly up to 3 cells wide. Unisereate rays are made up of upright cells only. Most rays have many rows of upright marginal cells.

Table 2.28 Anatomical characters of *Lonicera* spp. (Carpifoliaceae)

Character	*Lonicera periclymenum*	*Lonicera nigra/xylosteum*
TS		
Vessel arrangement	diffuse porous – dense throughout	Semi-ring porous (– diffuse porous) – sparse in late wood
Vessel grouping	solitary (clusters)	solitary (clusters)
Tyloses	present	present
Parenchyma	apotracheal diffuse	apotracheal diffuse
Ground tissue	± thin-walled throughout	thick-walled in late wood
Ring boundary	distinct	distinct
Ray/ring boundary	–	–
TLS		
Ray width	1–3(4)	1–3(4)
Ray height	8–15	8–15
Sheath cells	present	present
RLS		
Ray anatomy	heterogeneous type 1	heterogeneous type 1
Libriform fibres	absent	absent
Fibre tracheids	present	present
Vascular tracheids	absent	absent
Spiral thickening	fine	fine
Crystals	absent	absent
Perforation plates	simple (scalariform)	simple (scalariform)

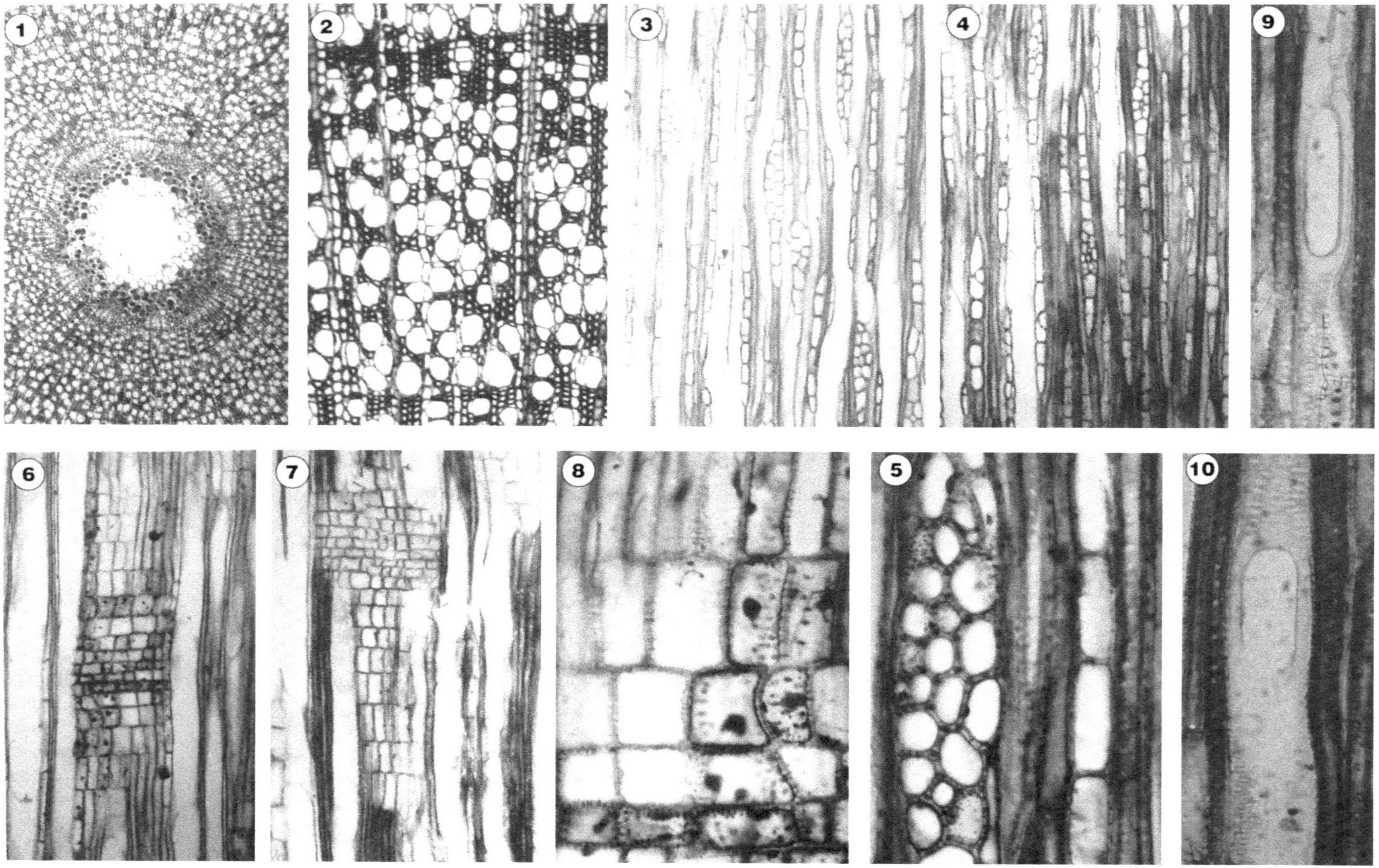

Plate 28 *Lonicera periclymenum* (1) TS pith and young wood (×28) (2) TS (×65) (3) TLS (×65) (4) TLS (×65) (5) TLS (×265) (6) RLS ray (×65) (7) RLS ray (×65) (8) RLS ray (×265) (9) scalariform perforation plate (×265) (10) scalariform perforation plate (×265)

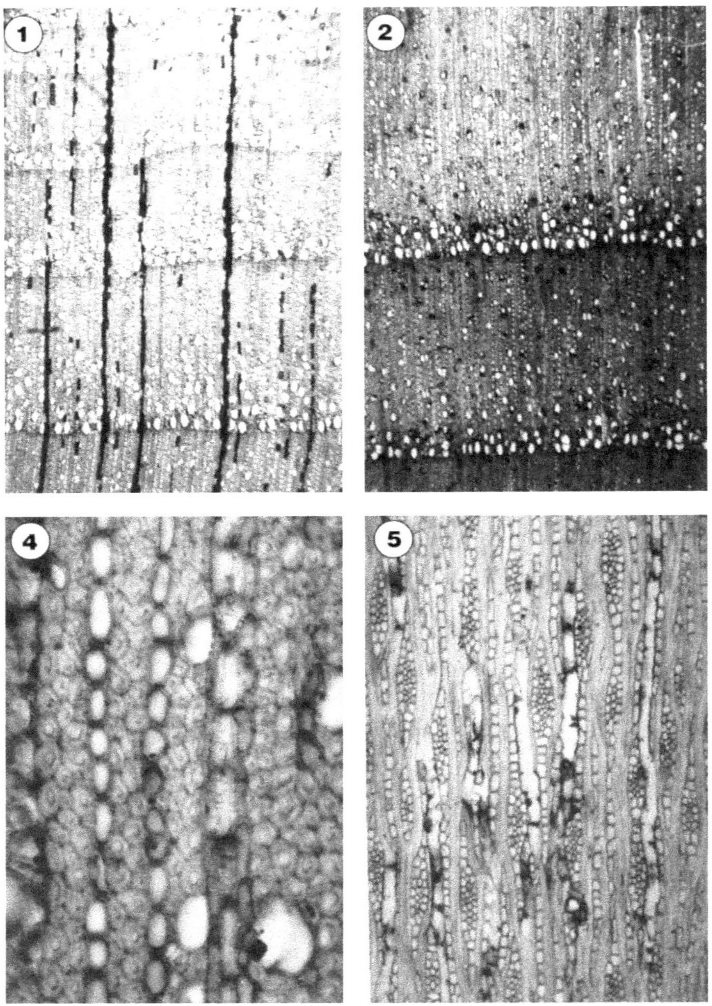

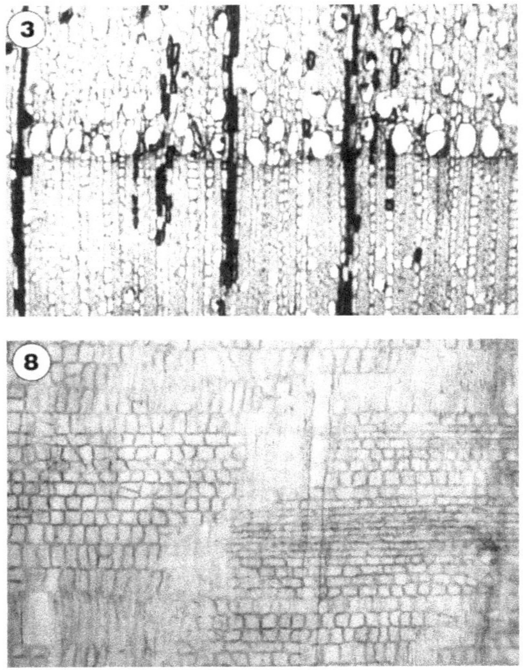

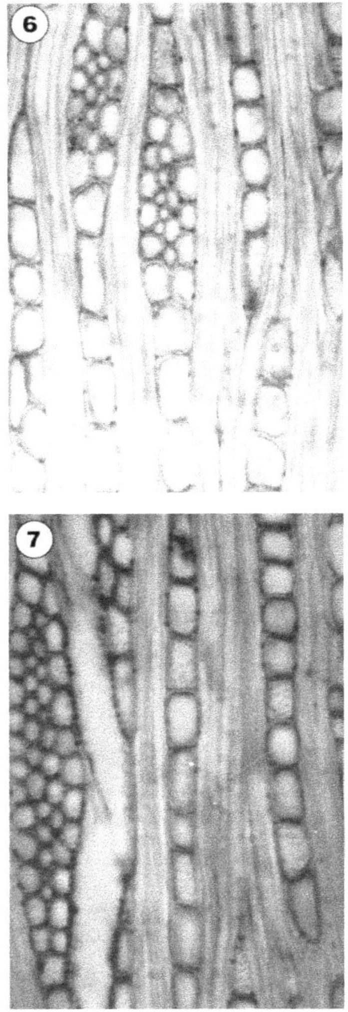

Plate 29 *Lonicera xylosteum* (1) TS (×28) (2) TS (×28) (3) TS (×65) (4) TS (×265) (5) TS (×65) (6) TLS (×265) (7) TLS (×265) (8) RLS (×65)

Unisereate rays present and rays up to 20 sereate

Fagus sylvatica – Beech (Fagaceae)

Plate 30

Fagus sylvatica is one of the most important and familiar central and western European large forest or woodland trees (see Map 58). It has always been important economically as both a structural wood and for its use in the manufacture in domestic and industrial artefacts.

Fagus sylvatica has a diffuse porous distribution of its vessels which gradually decrease in size over a year's growth. Parenchyma is apotracheal diffuse and in short tangential bands. Rays are mostly homogeneous, unisereate and up to 20 cells across or more, larger rays are up to 5 mm or more high. In TS rays, especially larger rays, are distended at the ring boundary. Both simple and scalariform perforation plates are present.

Fagus sylvatica is the only species with a diffuse porous distribution of mostly solitary vessels and large multisereate rays and as such it would be difficult to confuse it with other woods.

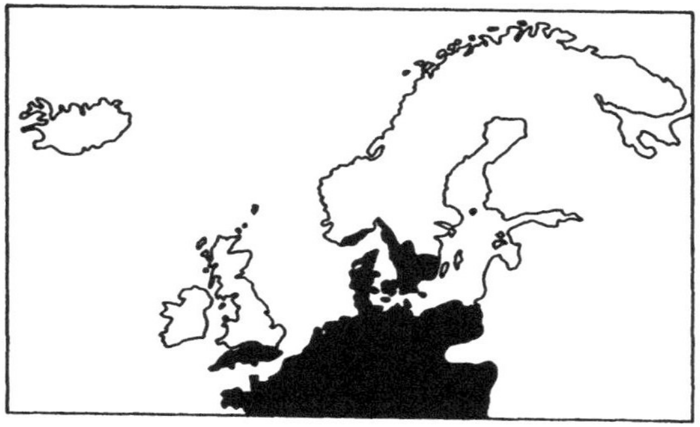

Map 58 Distribution of *Fagus sylvatica*

Table 2.29 Anatomical characters of *Fagus sylvatica* (Fagaceae)

Character	*Fagus Sylvatica*	Character	*Fagus Sylvatica*
TS		RLS	
Vessel arrangement	diffuse porous to semi-ring porous	Ray anatomy	homogeneous
Vessel grouping	± solitary	Libriform fibres	absent
Early–late wood transition	–	Fibre tracheids	present
Late wood vessels	–	Vasicentric tracheids	absent
Tyloses	present in early wood	Vascular tracheids	absent
Parenchyma	apotracheal diffuse, short tangential bands	Spiral thickening	absent
Ground tissue	thick-walled	Crystals	absent
Ring boundary	undulating	Perforation plates	simple (scalariform)
Ray/ring boundary	rays distended		
TLS			
Ray width	1, –20		
Ray height	–5 mm		
Sheath cells	absent		

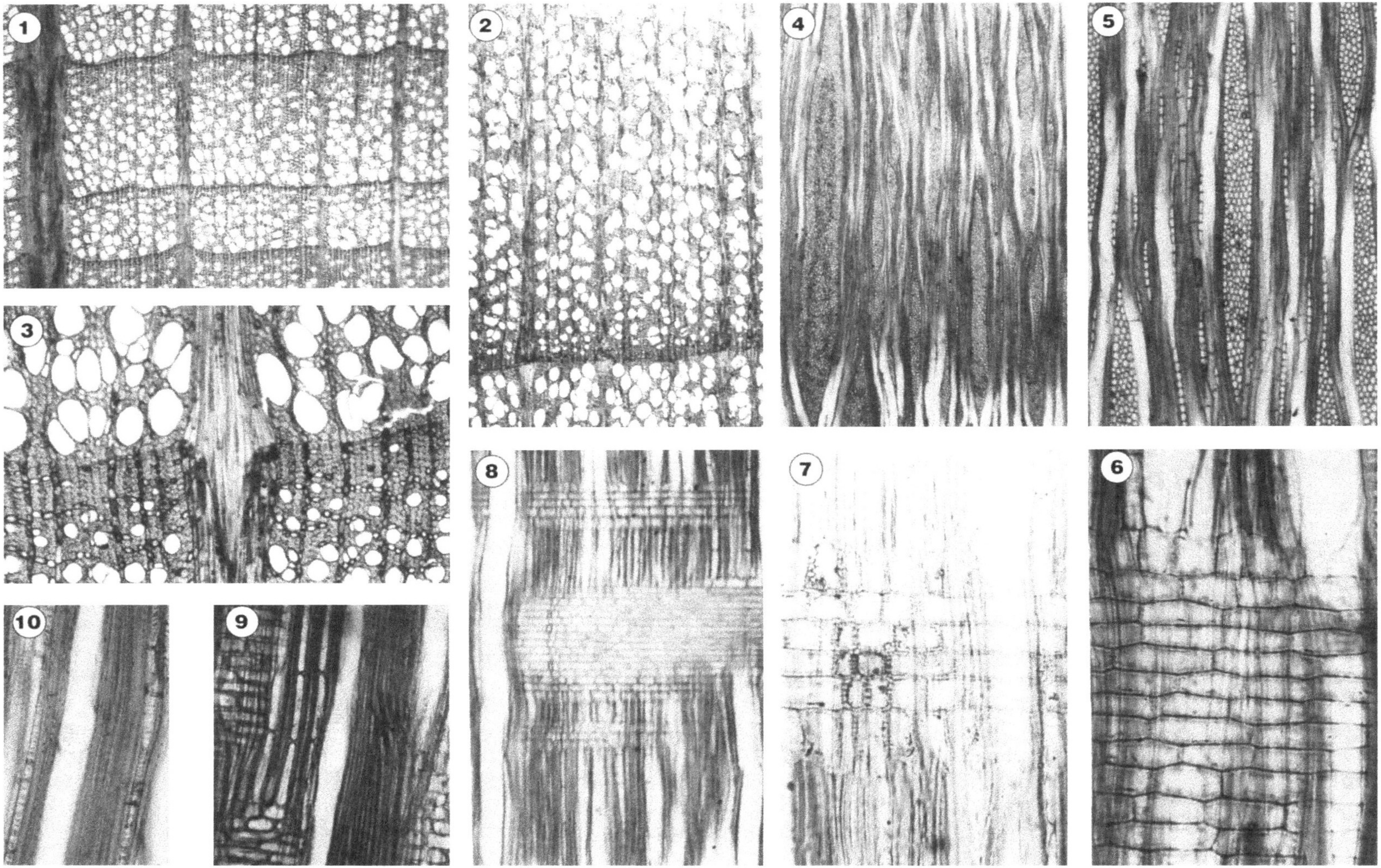

Plate 30 *Fagus sylvatica* (1) TS (×28) (2) TS (×65) (3) TS showing ray distended at ring boundary (×140) (4) TLS (×65) (5) TLS (×140) (6) RLS (×265) (7) RLS (×265) (8) RLS (×140) (9) RLS simple perforation plate (×140) (10) RLS scalariform perforation plate (×140)

Group 5 – Diffuse porous woods with solitary pores, pores in short radial files and/or clusters

This is a diverse and complex group within which are species groups that are difficult to separate (e.g. *Prunus, Acer,* Salicaceae). The group contains some economically important large trees such as lime and poplar as well as some of the less common smaller trees such as Alder buckthorn (*Frangula alnus*) and the Strawberry tree (*Arbuteus unedo*).

General anatomy

This group is defined by having vessels predominantly in clusters and short radial files as well as having solitary vessels. It is a large and diverse group which has been subdivided on ray size. The largest of these subdivisions, having both unisereate and small multisereate rays, is a complex group and is dealt with under *Frangula* at the end of this section.

Table 2.30 Anatomical characters of diffuse porous woods with vessels in clusters and short radial files

Taxon	Vessel grouping	Unisereate rays present	Largest ray width	Perforation plates	Spiral thickening present
Betula pendula, B. pubescens	solitary/short radial files	no	4	scalariform	no
Populus spp	solitary, short radial files/clusters	yes	1	simple	no
Salix spp	Solitary, short radial files/clusters	yes	1	simple	no
Rubus idaeus, R. fruticosus agg	solitary/clusters	yes	12	simple	no
Ficus carica	solitary/short radial files	yes	4	simple	yes
Tilia cordata, T. platyphyllos	solitary, short radial files/clusters	yes	4	simple	yes
Arbuteus unedo	Solitary, short radial files/clusters	yes	3	simple (sclfm)	yes
Prunus avium	solitary/clusters	yes	4	simple	yes
Prunus padus	solitary/clusters	yes	7	simple	yes
Prunus spinosa, P. domestica	solitary/clusters	yes	7	simple	yes
Acer campestre	solitary/short radial files	yes	3	simple	yes
Acer pseudoplatanis, A. platanoides	solitary/short radial files	yes	6	simple	yes
Juglans regia	solitary/short radial files	yes	5	simple	no
Frangula alnus	solitary, short radial files/clusters	yes	3	simple	yes (fine)

In addition to the taxa listed above, five ring porous taxa have late wood vessels distributed in clusters and short radial files as well as having solitary vessels. These are listed below with their most relevant other characteristics. All have simple perforation plates.

Both *Clematis vitalba* and *Vitis vinifera* have narrow stems with very tall rays and could not be confused with the diffuse porous taxa described in this group. The late wood of *Fraxinus, Hippophae* and *Morus* could be confused with these diffuse porous woods although for each of these the following characteristics will help with differentiation though the section of ring porous woods should be consulted.

- *Fraxinus excelsior*: very distinctive paratracheal parenchyma
- *Morus nigra*: homogeneous rays up to 7 cells wide and up to 60 cells high
- *Hippophae rhamnoides*: rays are narrow (unisereate and bisereate) and storied (seen in TLS)

Taxa descriptions

Group key

Unisereate rays absent	*Betula, Ficus*
Unisereate rays present	
• All rays unisereate	Salicaceae
• Unisereate and multiserteate rays present	
– Multisereate rays over 50 cells high	*Rubus*
– Multisereate rays under 50 cells high	*Tilia, Arbuteus, Ficus, Prunus, Acer, Juglans, Frangula*

Table 2.31 Anatomical characters of ring porous woods with clusters and short radial files

Taxa/character state	Late wood vessel grouping	Ray anatomy	Ray width	Ray height	Spiral thickening
Clematis vitalba	S, C	Ht 1,2,3	5-15+	10cm+	present
Hippophae rhamnoides	S, C	Hm (Ht 3)	1–2(–3)	5–15	present
Fraxinus excelsior	S, SRF, C	Ht 1,2,3	(1–)2–3	10–20	absent
Morus nigra	S, SRF, C	Ht 2,3	4–7	–60	present
Vitis vinifera	S, SRF (C)	Hm (Ht 2, 3)	5–20	–2mm	absent

Unisereate rays absent

Betula pendula – Silver birch

Betula pubescens – Downy birch (Betulaceae)

Plate 31

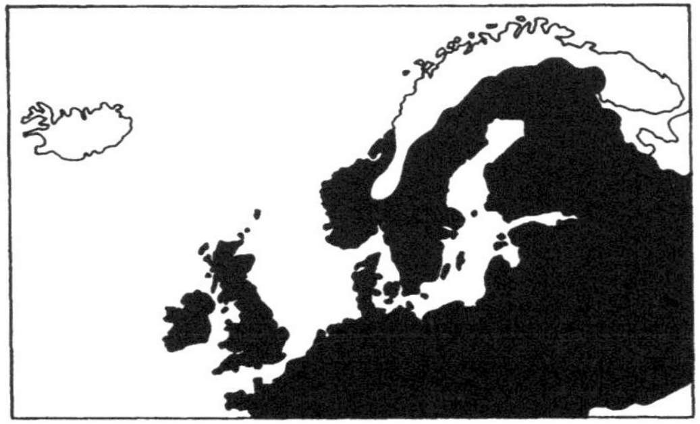

Map 59 Distribution of *Betula pendula*

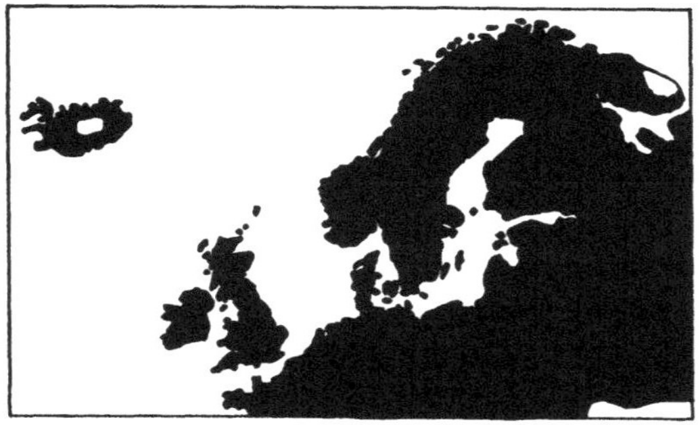

Map 60 Distribution of *Betula pubescens*

The birches are some of the most familiar and common of the European trees, being a major component of the boreal woodland with pine and spruce. It is of economic importance as a structural element and in artefact manufacture. Both species have broad distributions in northern Europe and Scandinavia. *Betula pubescens* is more widespread in northern Scandinavia and is present in Iceland (see Maps 59 and 60).

The two species cannot be differentiated. Both have a diffuse porous distribution of vessels with short radial files of 2 to 3, rarely 4, vessels as well as solitary vessels. Vessel diameter does not decrease much over a year's growth. The ring boundary is marked by several rows of terminal flattened cells. Rays are slightly distended at the ring boundary. Rays are mostly homogeneous, bisereate to 4-sereate and up to 15 cells high. Aggregate rays may be present but are rare. Perforation plates are scalariform with up to 25 narrow bars.

If aggregate rays are present *Betula* may be confused with *Alnus* since both have very similar scalariform perforation plates, however, *Betula* species have larger rays outside of the aggregate rays.

Table 2.32 Anatomical characters of *Betula* (Betulaceae)

Character	*Betula*	Character	*Betula*	Character	*Betula*
TS		TLS		RLS	
Vessel arrangement	diffuse porous	Ray width	2–4(–5)	Ray anatomy	homogeneous (heterogeneous type 3)
Vessel grouping	solitary, short radial files	Ray width (aggregate rays)	–	Libriform fibres	present
Tyloses	–	Ray height	10–15	Fibre tracheids	absent
Parenchyma	apotracheal diffuse	Sheath cells	absent	Vascular tracheids	absent
Ground tissue	–			Spiral thickening	absent
Ring boundary	smooth			Perforation plates	scalariform (10–25)
Ray/ring boundary	very distinct				
Aggregate rays	(very rare)				

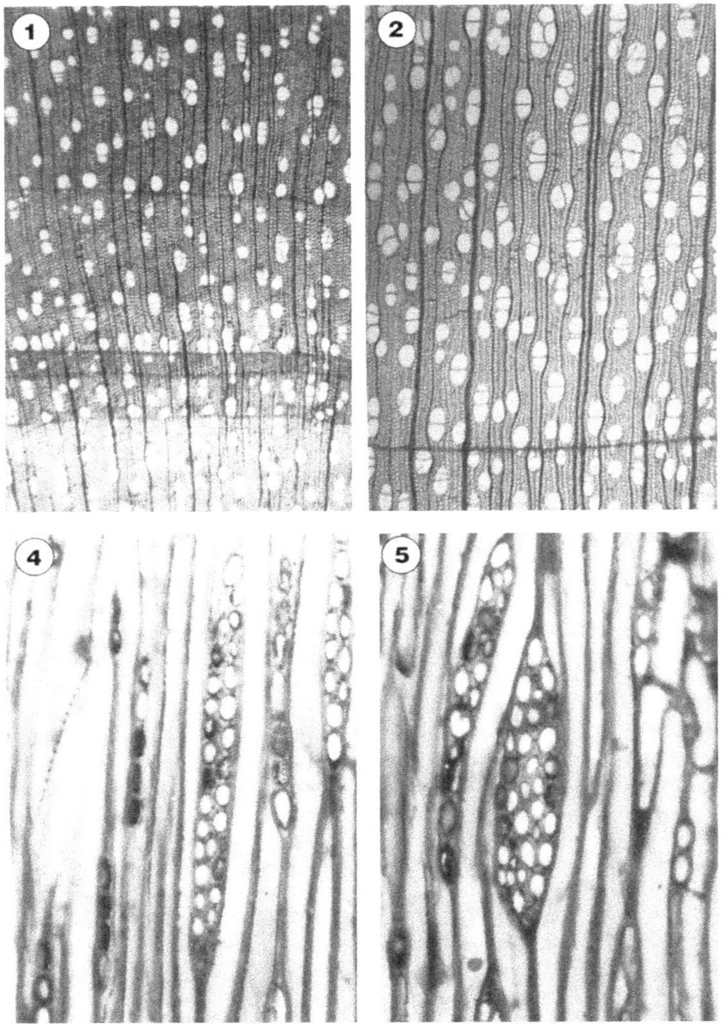

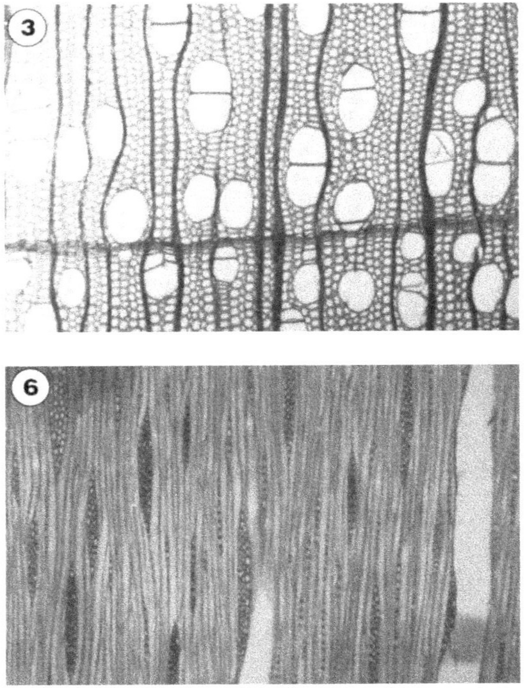

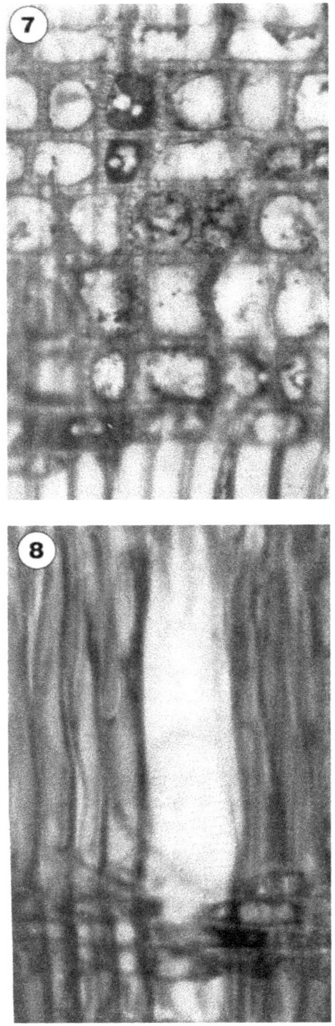

Plate 31 *Betula pendula* (1) TS (×28) (2) TS (×65) (3) TS (×140) (4) TLS (×265) (5) TLS (×265) (6) TLS (×140) (7) RLS ray (×265) (8) RLS scalariform perforation plate (×265).

Unisereate rays present, All rays unisereate

Salicaceae – *Salix* – the willows and *Populus* – the poplars

Plate 32 and Plate 33

The two genera in this large family are very similar anatomically and not always easy to separate. It is an important group in northern Europe both economically and ecologically. The poplars are mostly large trees whereas the willows display much greater variation in habit from being large trees to procumbent shrubs. Distributions of the following are noted.

Polulus alba (see Map 61)
Polulus nigra (see Map 62)
Polulus canescens (see Map 63)
Polulus tremula (see Map 64)
Salix alba (see Map 65)
Salix arbuscula (see Map 66)
Salix atrocinera (see Map 67)
Salix aurita (see Map 68)
Salix artica (see Map 69)
Salix acutifolia (see Map 70)
Salix bicolor (see Map 71)
Salix burjatica (see Map 72)
Salix capraea (see Map 73)
Salix cinerea (see Map 74)
Salix daphnoides (see Map 75)
Salix eleagnos (see Map 76)
Salix fragilis (see Map 77)
Salix glauca (see Map 78)
Salix hastata (see Map 79)
Salix herbacea (see Map 80)
Salix hibernica (see Map 81)
Salix lantana (see Map 82)
Salix lapponum (see Map 83)
Salix myrsinifolia (see Map 84)
Salix myrsinites (see Map 85)
Salix myrtiloides (see Map 86)
Salix pentandra (see Map 87)
Salix phylicifolia (see Map 88)
Salix polaris (see Map 89)
Salix purpurea (see Map 90)
Salix pyrofolia (see Map 91)
Salix repens (see Map 92)
Salix reptans (see Map 93)
Salix reticulata (see Map 94)
Salix rosmarinifolia (see Map 95)
Salix starkeana (see Map 96)
Salix triandra (see Map 97)
Salix viminalis (see Map 98)
Salix xerophyla (see Map 99)

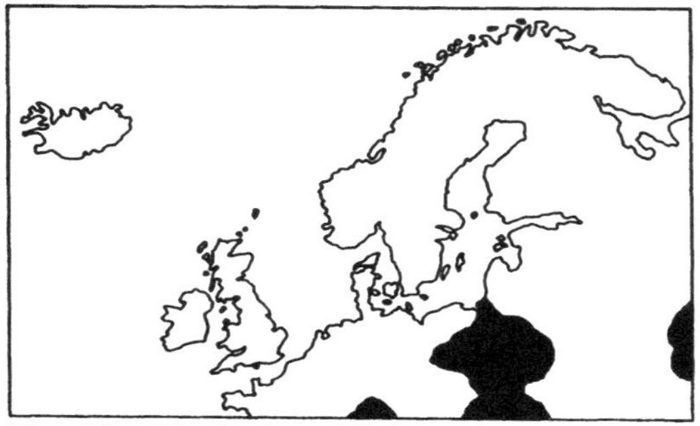
Map 61 Distribution of *Polulus alba*

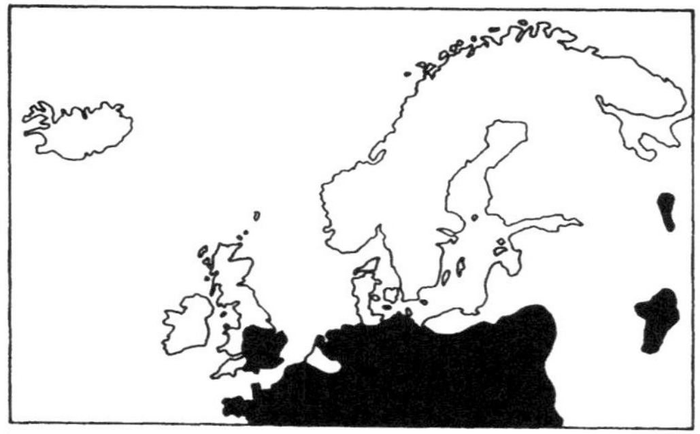
Map 62 Distribution of *Polulus nigra*

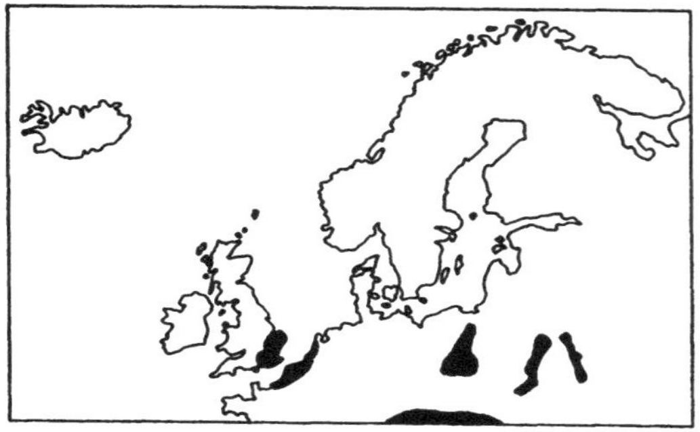
Map 63 Distribution of *Polulus canescens*

The Salicaceae has a diffuse porous distribution of vessels distributed mostly in short radial files, clusters and as solitary vessels. Vessels on the whole do not decrease much in

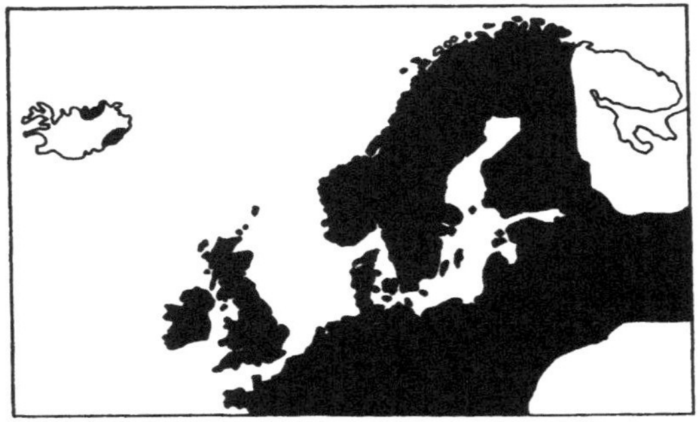

Map 64 Distribution of *Polulus tremula*

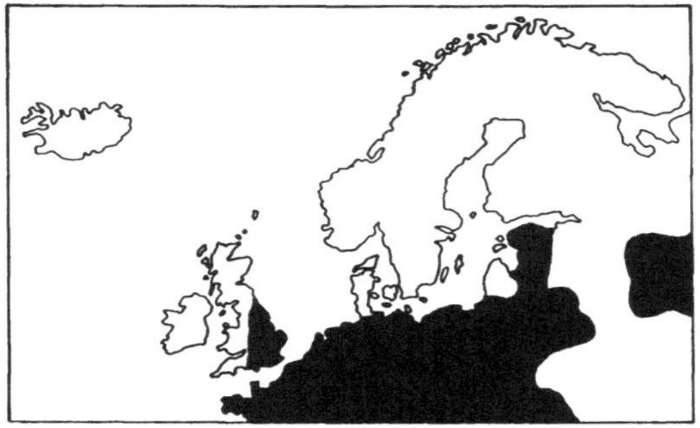

Map 65 Distribution of *Salix alba*

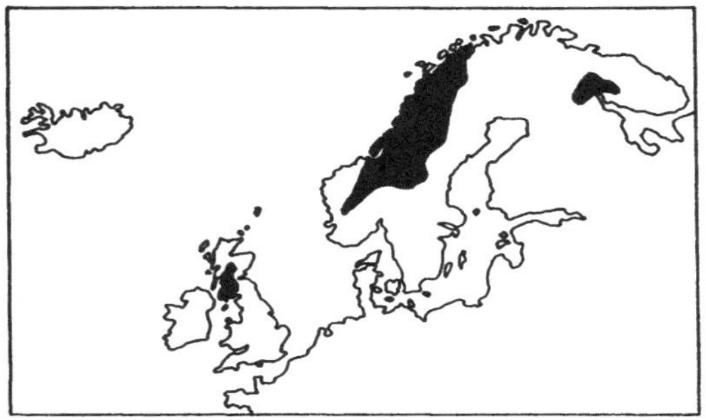

Map 66 Distribution of *Salix arbuscula*

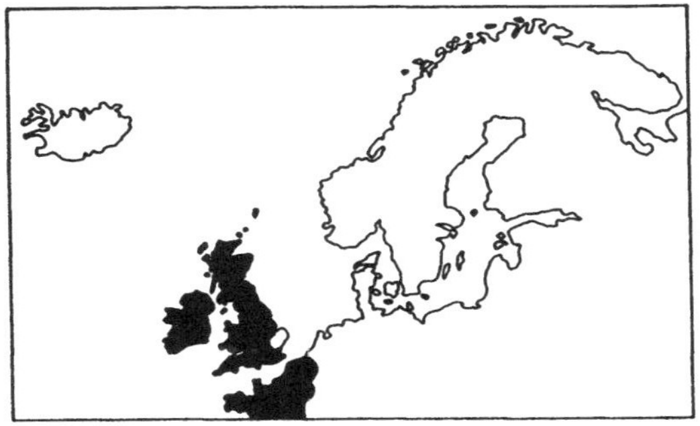

Map 67 Distribution of *Salix atrocinera*

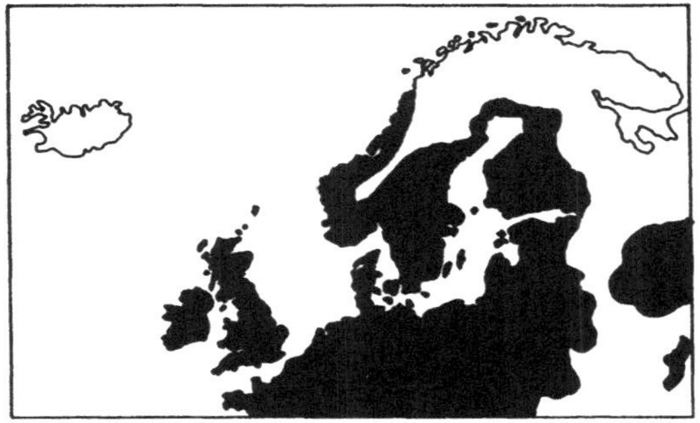

Map 68 Distribution of *Salix aurita*

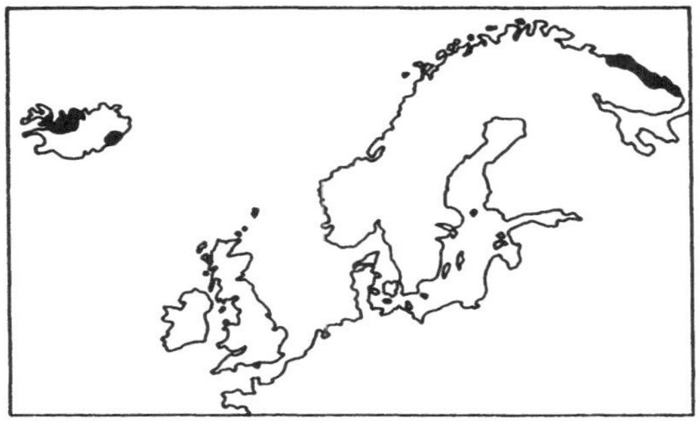

Map 69 Distribution of *Salix artica*

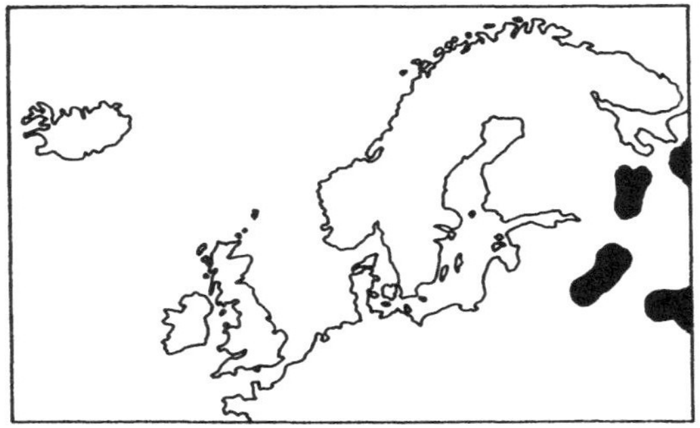

Map 70 Distribution of *Salix acutifolia*

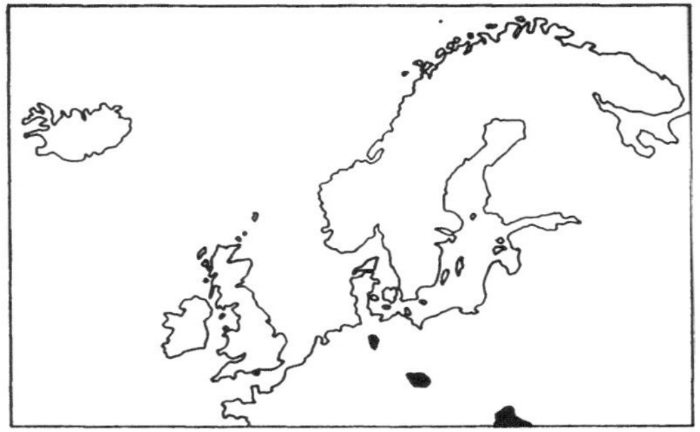

Map 71 Distribution of *Salix bicolor*

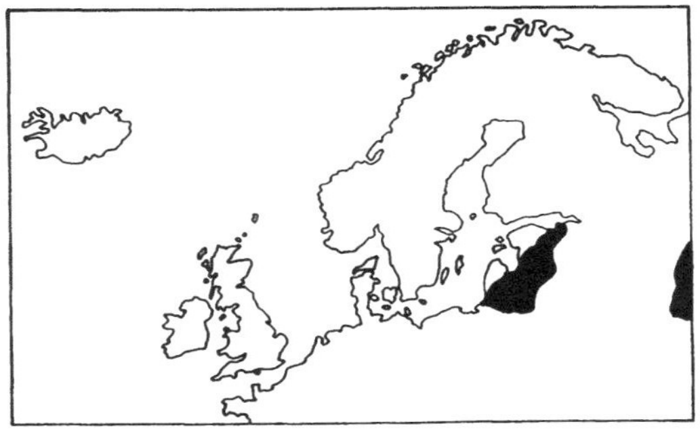

Map 72 Distribution of *Salix burjatica*

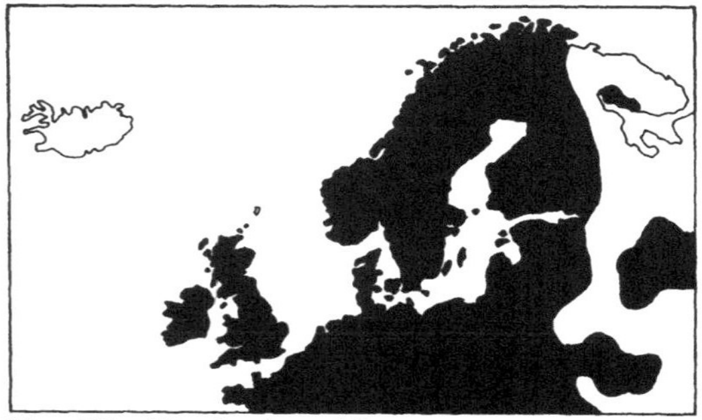

Map 73 Distribution of *Salix capraea*

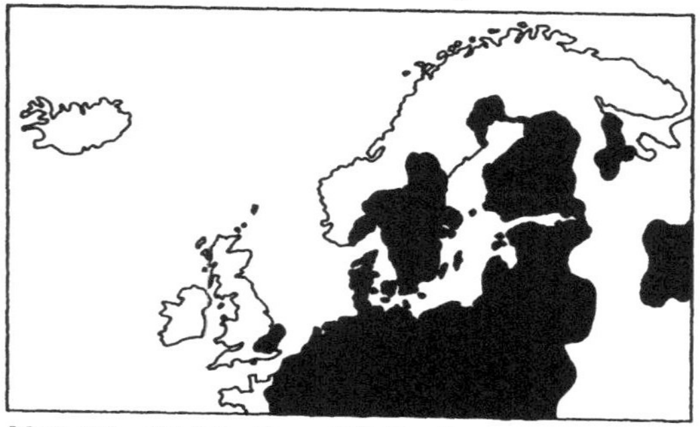

Map 74 Distribution of *Salix cinerea*

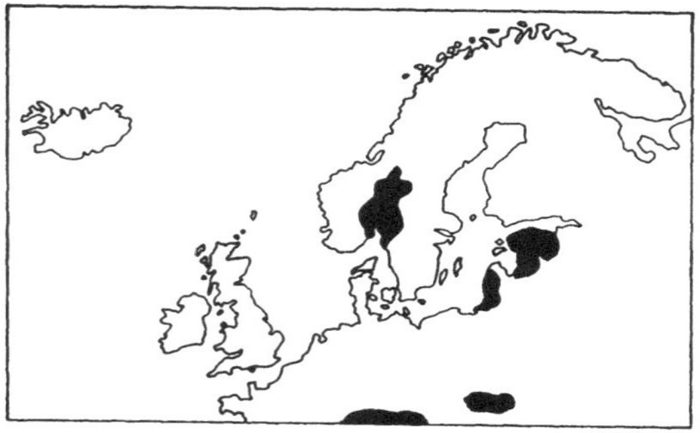

Map 75 Distribution of *Salix daphnoides*

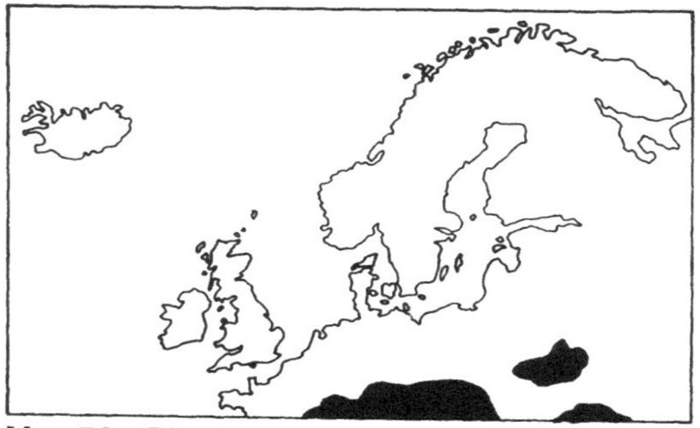

Map 76 Distribution of *Salix eleagnos*

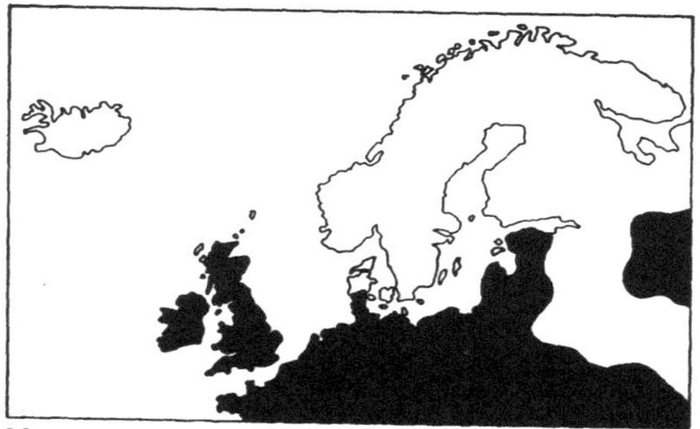

Map 77 Distribution of *Salix fragilis*

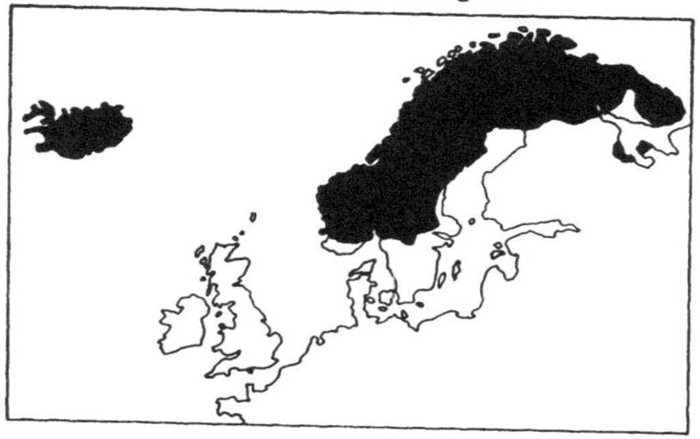

Map 78 Distribution of *Salix glauca*

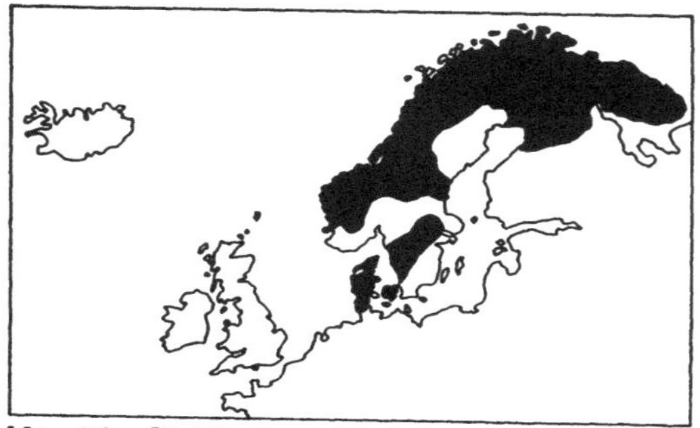

Map 79 Distribution of *Salix hastata*

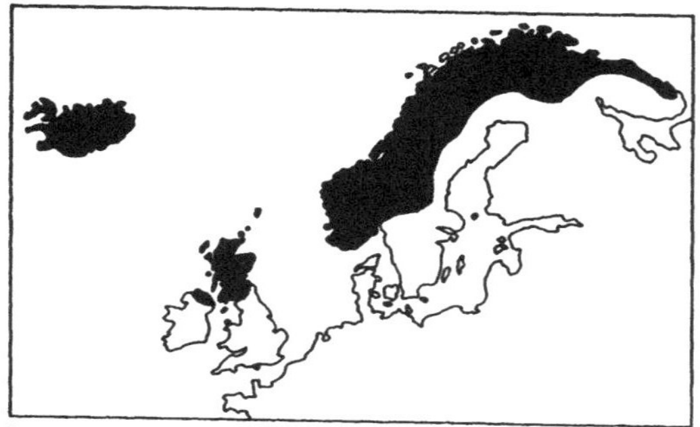

Map 80 Distribution of *Salix herbacea*

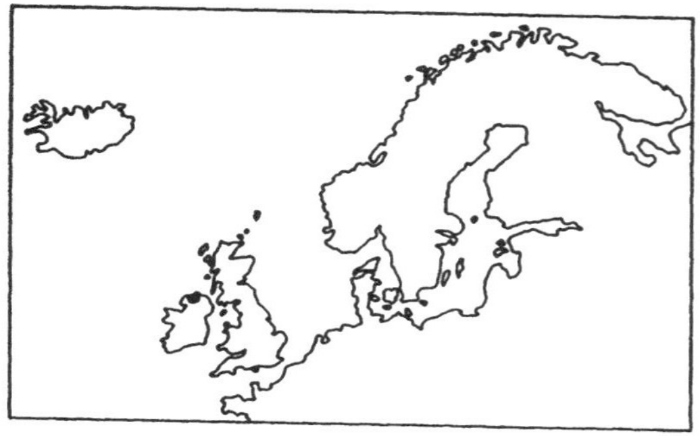

Map 81 Distribution of *Salix hibernica*

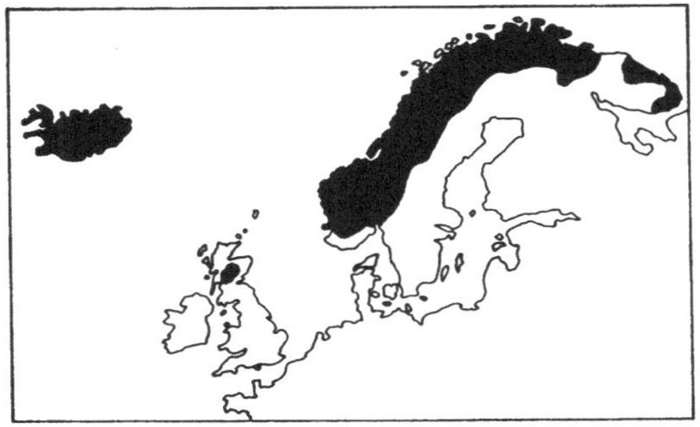

Map 82 Distribution of *Salix lantana*

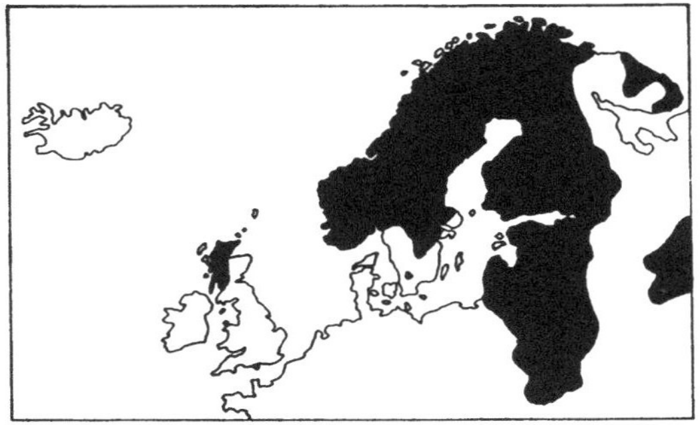

Map 83 Distribution of *Salix lapponum*

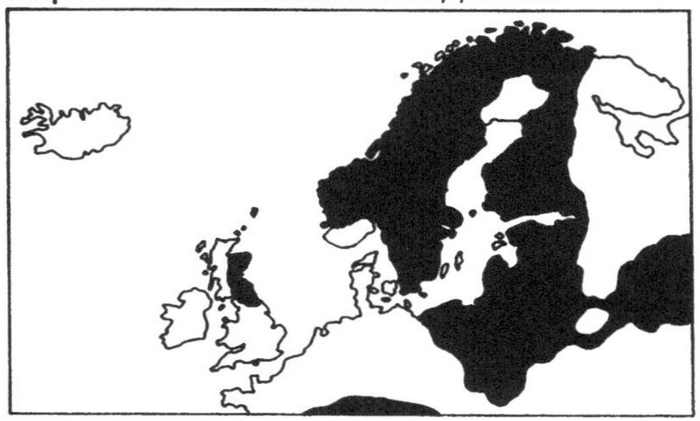

Map 84 Distribution of *Salix myrsinifolia*

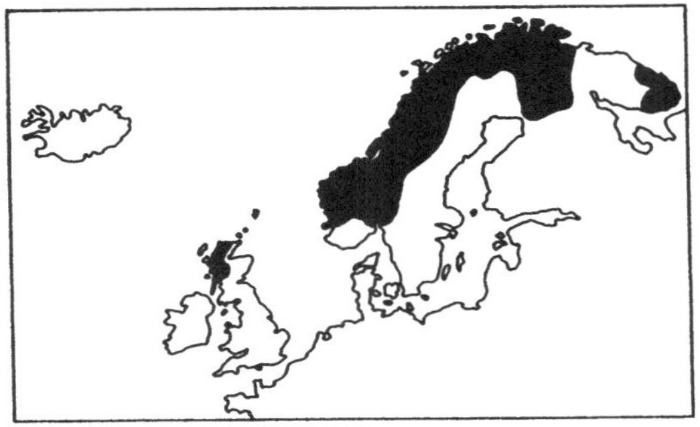

Map 85 Distribution of *Salix myrsinites*

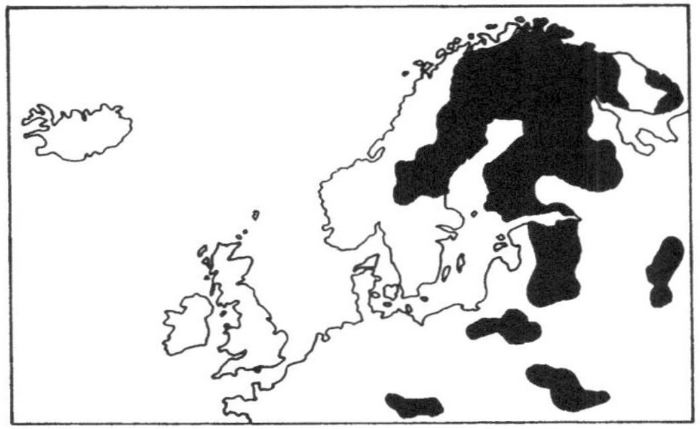

Map 86 Distribution of *Salix myrtiloides*

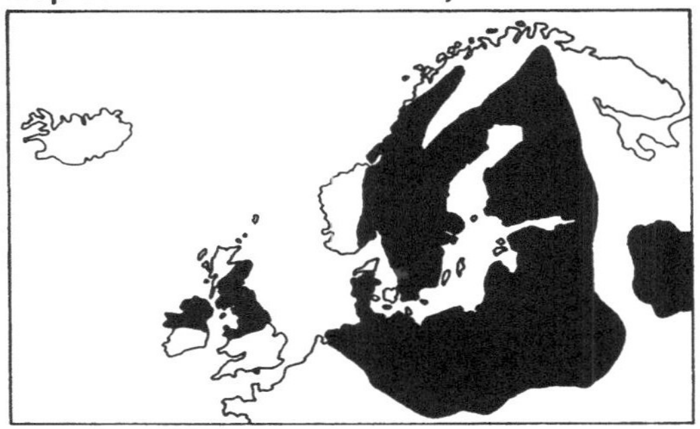

Map 87 Distribution of *Salix pentandra*

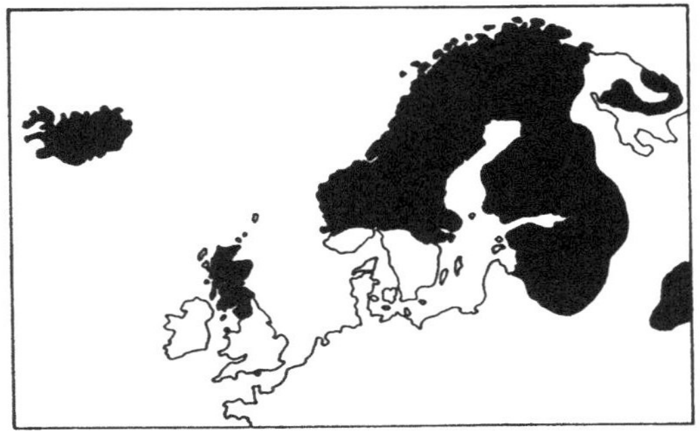

Map 88 Distribution of *Salix phylicifolia*

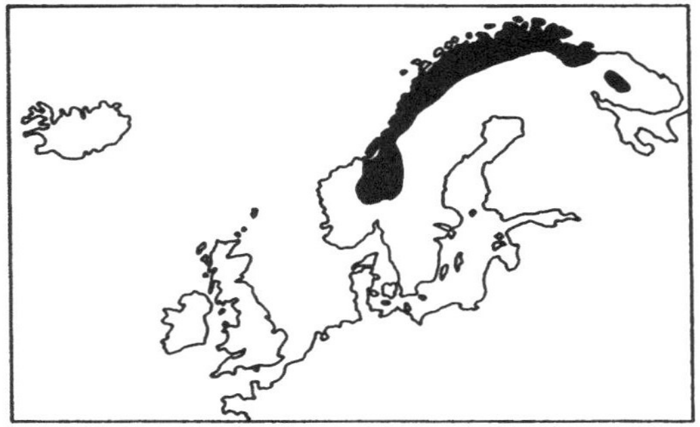

Map 89 Distribution of *Salix polaris*

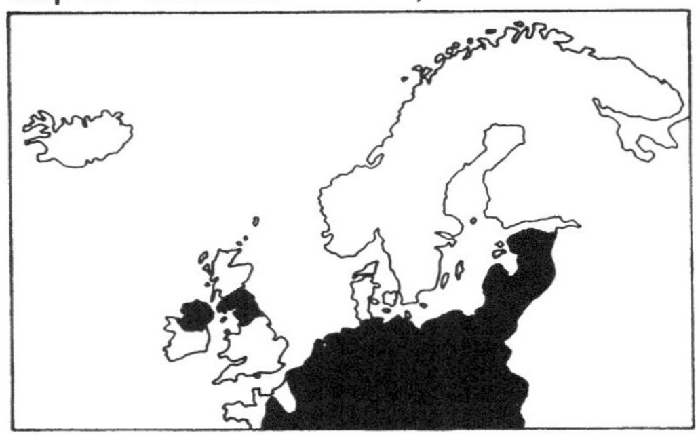

Map 90 Distribution of *Salix purpurea*

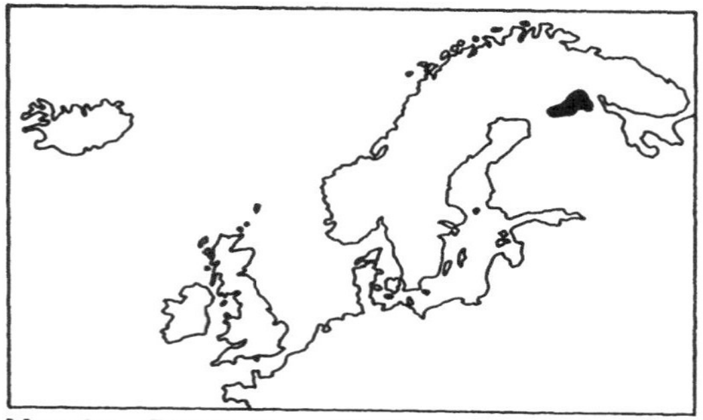

Map 91 Distribution of *Salix pyrofolia*

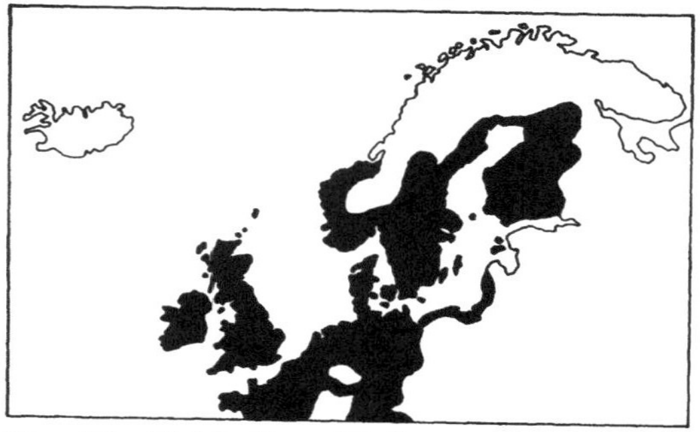

Map 92 Distribution of *Salix repens*

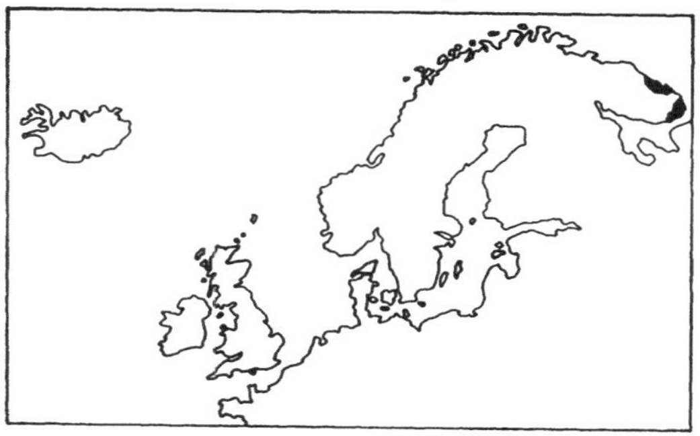

Map 93 Distribution of *Salix reptans*

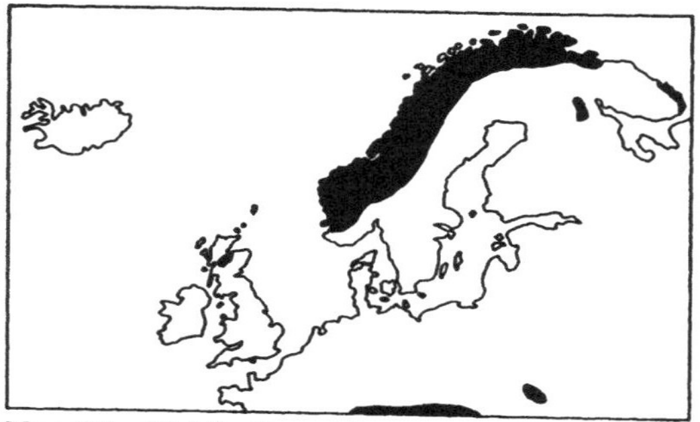

Map 94 Distribution of *Salix reticulata*

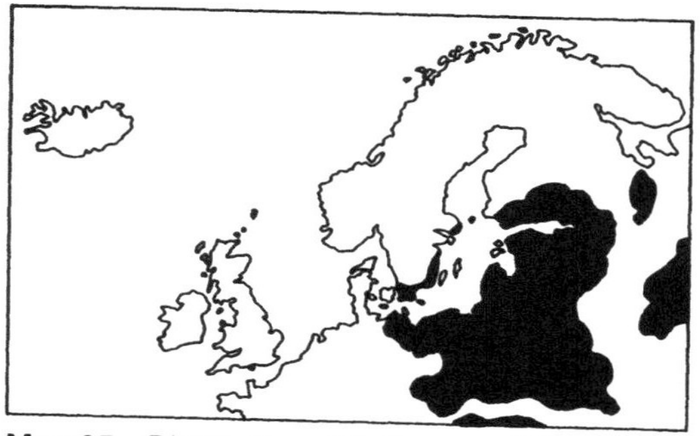

Map 95 Distribution of *Salix rosmarinifolia*

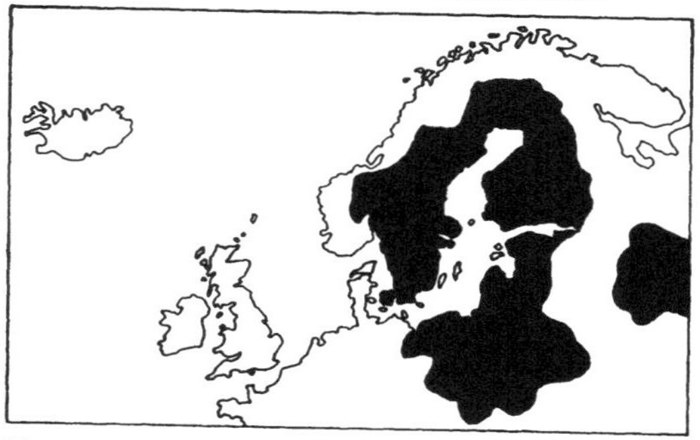

Map 96 Distribution of *Salix starkeana*

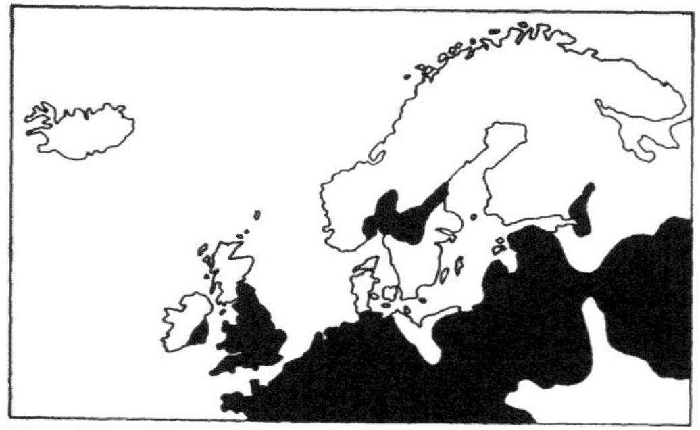

Map 97 Distribution of *Salix triandra*

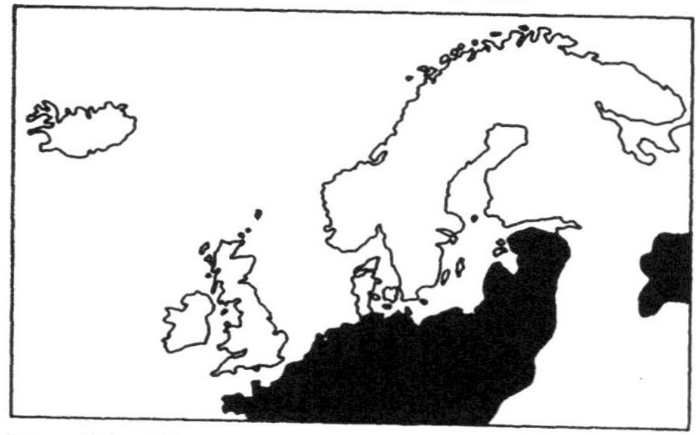

Map 98 Distribution of *Salix viminalis*

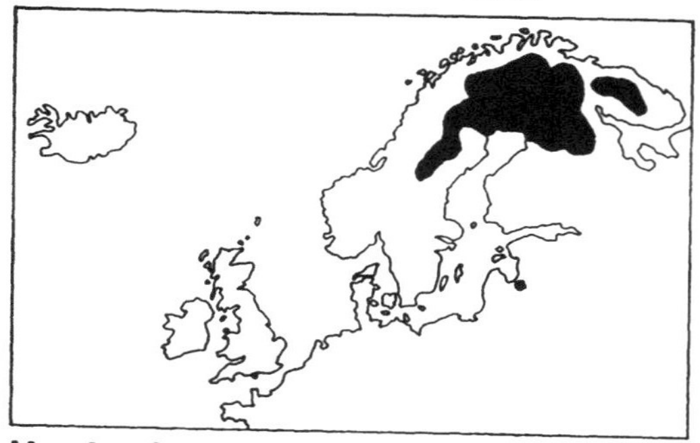

Map 99 Distribution of *Salix xerophyla*

size over a year's growth and so ring boundaries may be rather indistinct. Parenchyma is apotracheal diffuse and may also be banded terminal. Both homogeneous and heterogeneous rays are present (see below – differentiating between *Salix* and *Populus*). All rays are unisereate and up to 15 cells high. Perforation plates are simple.

Differentiating between Salix *and* Populus

The two genera are very similar in their anatomy with the exception of the predominance of heterogeneous rays, with a single row of square cells in *Salix* and homogeneous rays in *Populus*. However there are heterogeneous rays present in *Populus* and homogeneous rays present in *Salix*. In most cases the two genera cannot be separated, in no case can they be separated with certainty. Schweingruber (1990) suggests that while the tree willows (e.g. *S. alba*, *S. cinerea*) cannot be easily separated from the poplars, the bush willows (e.g. *S. arbuscula*) and prostrate willows (e.g. *S. reticulata*) may be separable from each other, but that more work needs to be done.

Table 2.33 Anatomical characters of the Salicaceae

Character	*Salix*	*Populus*
TS		
Vessel arrangement	diffuse porous (semi-ring porous)	diffuse porous (semi-ring porous)
Vessel grouping	Solitary, clusters, short radial files	solitary, clusters, short radial files
Tyloses	absent	absent
Parenchyma	apotracheal diffuse, banded terminal	apotracheal diffuse, banded terminal
Ground tissue	–	–
Ring boundary	variable	variable
Ray/ring boundary	–	–
TLS		
Ray width	1	1
Ray height	10–15	10–15
Sheath cells	absent	absent
RLS		
Ray anatomy	heterogeneous type 3 (homogeneous)	homogeneous (heterogeneous type 3)
Libriform fibres	present	present
Fibre tracheids	absent	absent
Vascular tracheids	absent	absent
Spiral thickening	absent	absent
Perforation plates	simple	simple

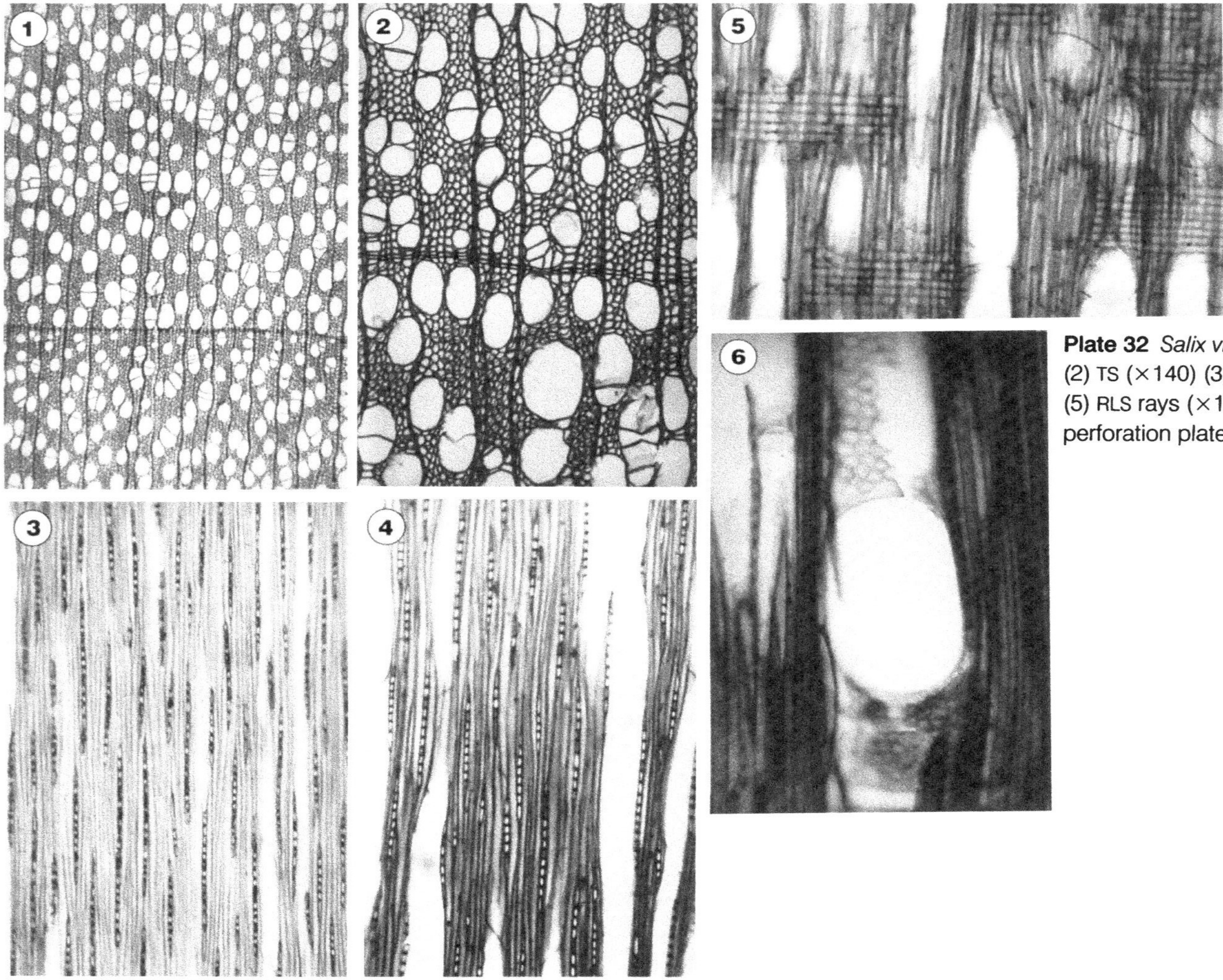

Plate 32 *Salix viminalis* (1) TS (×28) (2) TS (×140) (3) TLS (×65) (4) TLS (×140) (5) RLS rays (×140) (6) RLS simple perforation plates (×265).

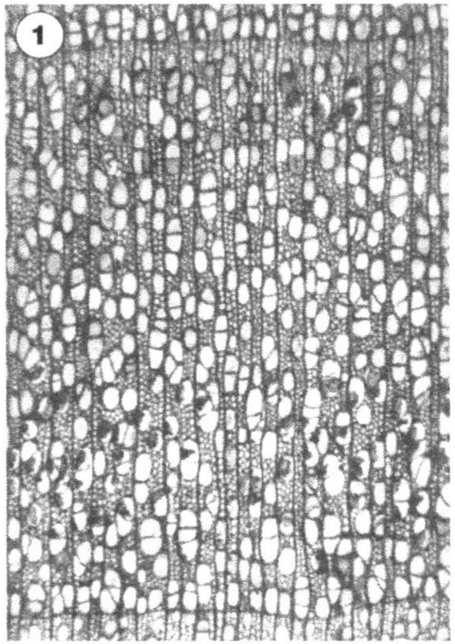

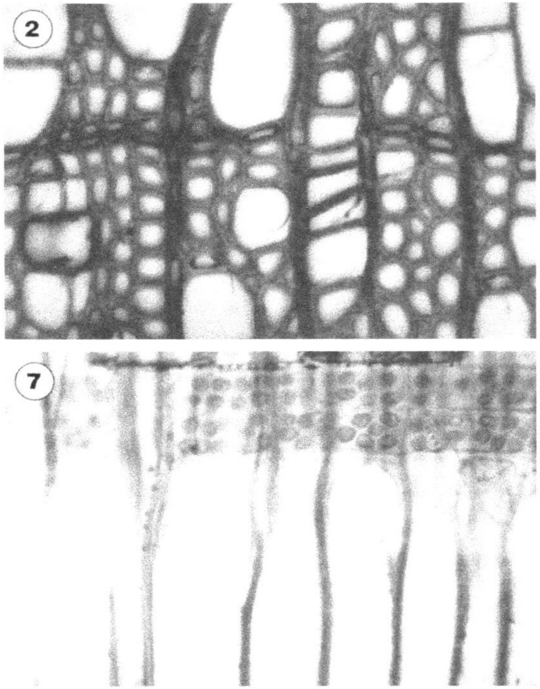

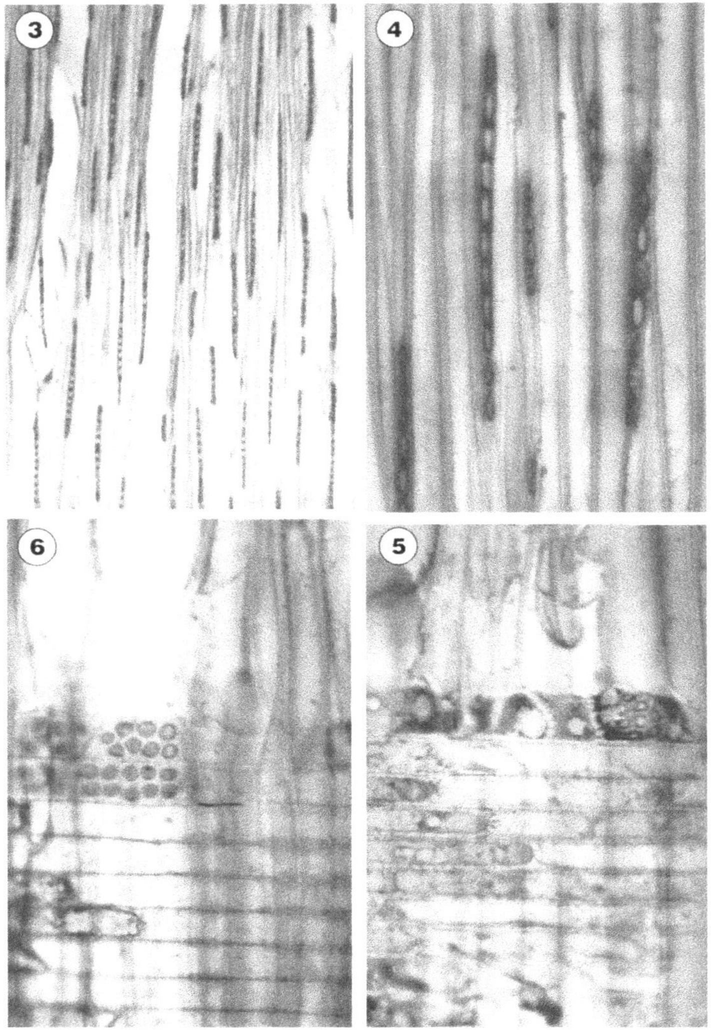

Plate 33 *Populus nigra* (1) TS (×28)
(2) TS (×265) (3) TLS (×140) (4) TLS (×140)
(5) RLS ray (×265) (6) RLS ray (×265)
(7) RLS simple perforation plates (×265)

Unisereate and multisereate rays present, Multisereate rays over 50 cells high

Rubus fruticosus agg. – Blackberry

Rubus idaeus – Raspberry (Rosoideae, Rosaceae)

Plate 34

The two *Rubus* taxa are both scramblers, usually, at least partially, using the support of other plants to grow, though this is a feature more common in *R. fruticosus* than *R. idaeus*. *Rubus fruticosus* is distributed over much of western northern Europe and into Scandinavia, its distribution further east is undocumented (see Map 100). *Rubus ideaus* is distributed much more broadly across northern Europe and into Scandinavia (see Map 101).

Neither of these taxa produce much wood: *R. idaeus* woody shoots are alive for two season's growth before dying, *R. fruticosus* woody shoots may live for several years. The two taxa cannot be separated on the basis of their wood anatomy.

These two *Rubus* species have vessels that are distributed in small clusters or are solitary. Ground tissue is thick-walled and parenchyma is apotracheal diffuse and rather sparse. Growth ring boundaries are rather indistinct. Unisereate and multisereate rays are present. Unisereate rays are composed entirely of upright cells, 10 to 12 cells high; multisereate rays are heterogeneous with numerous rows of upright cells, up to 12 cells wide, though somewhat narrower in *R. idaeus* and 10 mm high. Sheath cells are present in the multisereate rays. Perforation plates are simple.

The presence of such large rays and limited width of growth should prevent confusion of *Rubus* with other woods.

Table 2.34 Anatomical characters of *Rubus fruticosus* and *R. idaeus* (Rosoideae – Rosaceae)

Character	*Rubus*
TS	
Vessel arrangement	diffuse porous
Vessel grouping	solitary, clusters
Tyloses	absent
Parenchyma	apotracheal diffuse
Ground tissue	thick-walled
Ring boundary	indistinct, undulating
Ray/ring boundary	–
TLS	
Ray width	1, 4–12
Ray height	12, –10 mm
Sheath cells	present
RLS	
Ray anatomy	homogeneous, heterogeneous type 1
Libriform fibres	absent
Fibre tracheids	present
Vascular tracheids	absent
Spiral thickening	absent
Crystals	absent
Perforation plates	simple

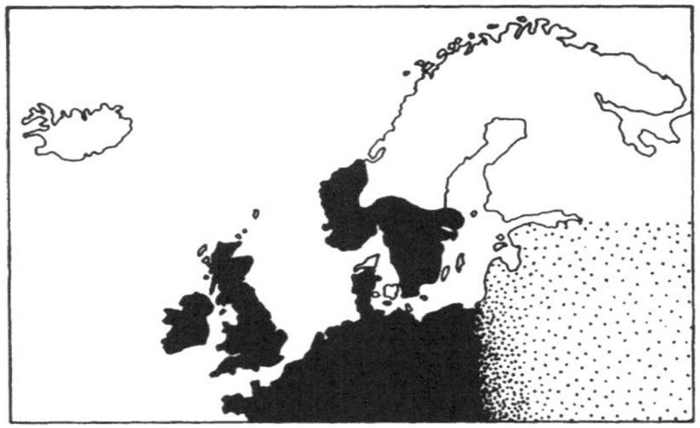

Map 100 Distribution of *Rubus fruticosus*

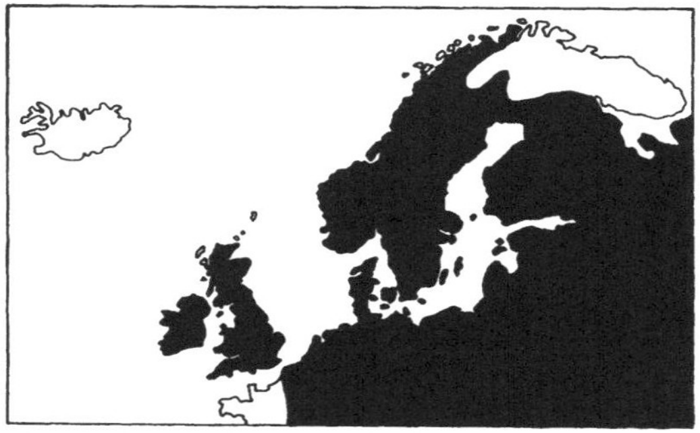

Map 101 Distribution of *Rubus idaeus*

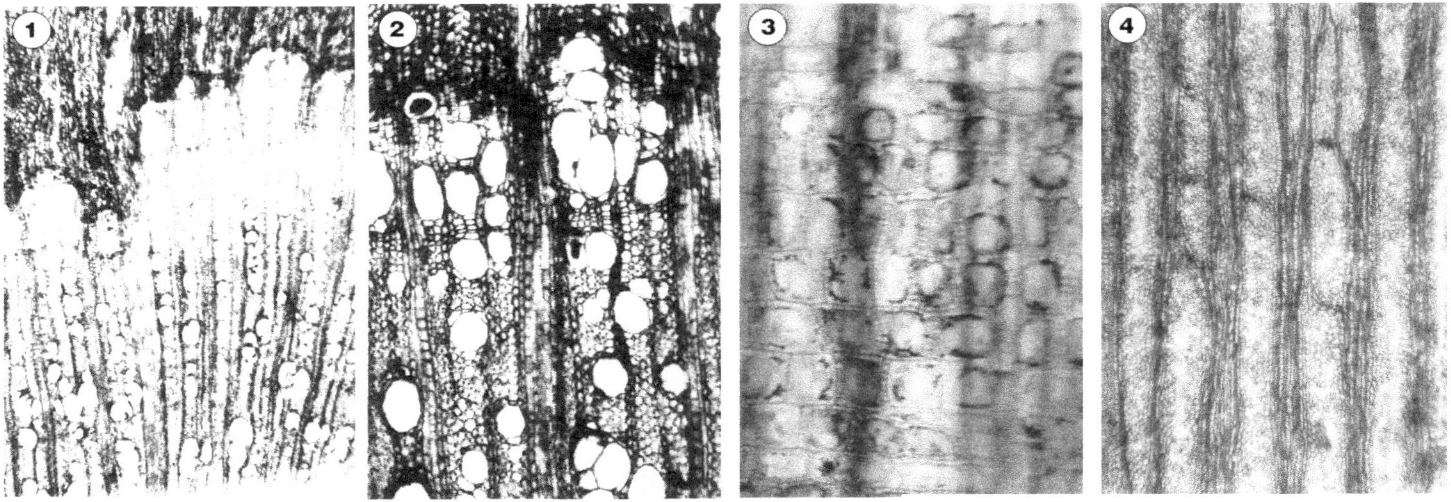

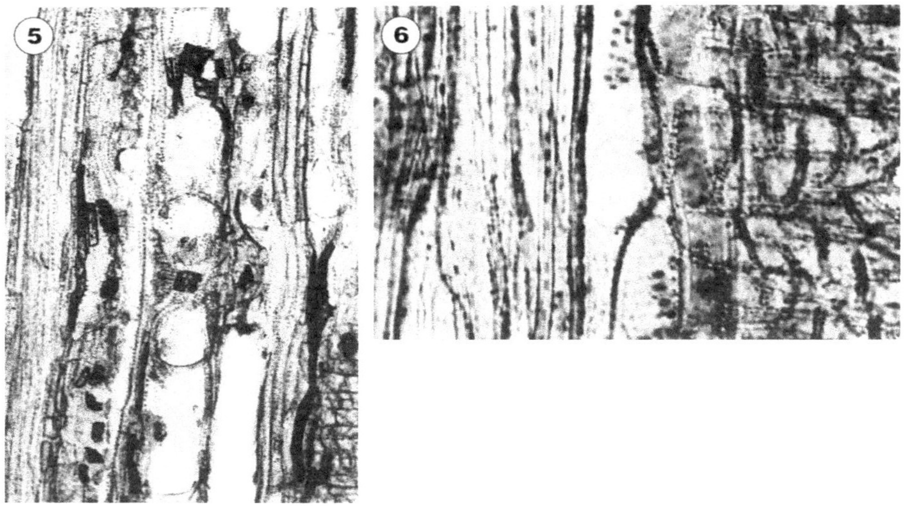

Plate 34 *Rubus fruticosus* agg. (1) TS (×28) (2) TS (×65) (3) RLS ray (×265) (4) TLS (×65) (5) RLS simple perforation plates (×265) (6) RLS simple perforation plates (×265).

Multisereate rays under 50 cells high

Tilia cordata – Small-leaved lime

Tilia platyphyllos – Large-leaved lime (Tiliaceae)

Plate 35

Both species of lime are large, ecologically and economically important trees distributed throughout northern Europe, though *T. cordata* alone is present in Scandinavia (see Maps 102 and 103).

Tilia has diffuse porous wood with vessels distributed in short radial files and clusters alongside solitary vessels. Ring boundaries are made distinct by the presence of terminal flattened cells. Ground tissue is distinctly thin-walled and rays are distended at the ring boundaries. Short tangential bands of parenchyma are present. Unisereate rays are composed of upright cells only and up to 7 cell high. Multisereate rays are homogeneous and heterogeneous with one row of upright cells, up to 4 sereate and 20 cells high, though more commonly shorter. Spiral thickening is very distinct. Perforation plates are simple.

The characteristics of thin-walled ground tissue, spiral thickening and unisereate rays with upright cells only make *Tilia* difficult to confuse with other diffuse porous woods.

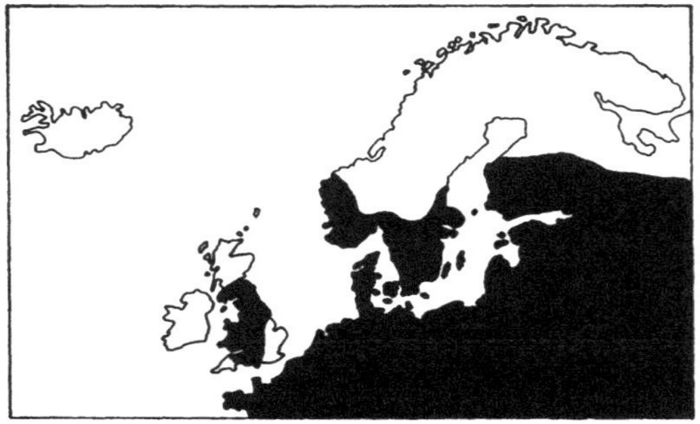

Map 102 Distribution of *Tilia cordata*

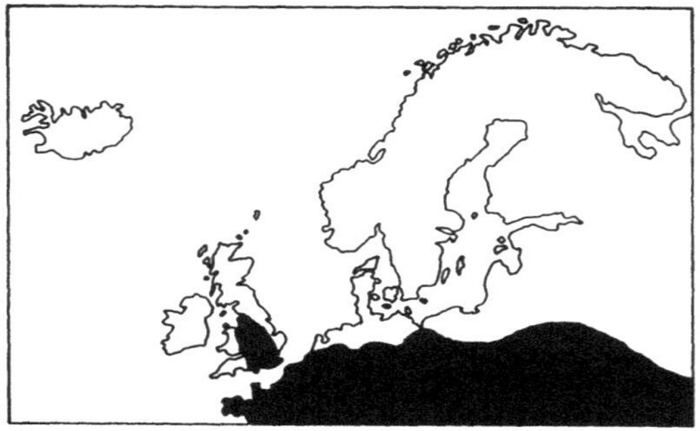

Map 103 Distribution of *Tilia platyphyllos*

Table 2.35 Anatomical characters of *Tilia* (Tiliaceae)

Character	*Tilia*	Character	*Tilia*	Character	*Tilia*
TS		TLS		RLS	
Vessel arrangement	diffuse porous (semi-ring porous)	Ray width	1, 2–4+	Ray anatomy	homogeneous, heterogeneous type 3
Vessel grouping	solitary short radial files, clusters	Ray height	5–10 (–20)		
Tyloses	absent	Sheath cells	absent	Libriform fibres	present
Parenchyma	apotracheal short tangential bands, (terminal)			Fibre tracheids	(present)
				Vascular tracheids	absent
Ground tissue	thin-walled			Spiral thickening	present
Ring boundary	distinct			Perforation plates	simple
Ray/ring boundary	rays distended				

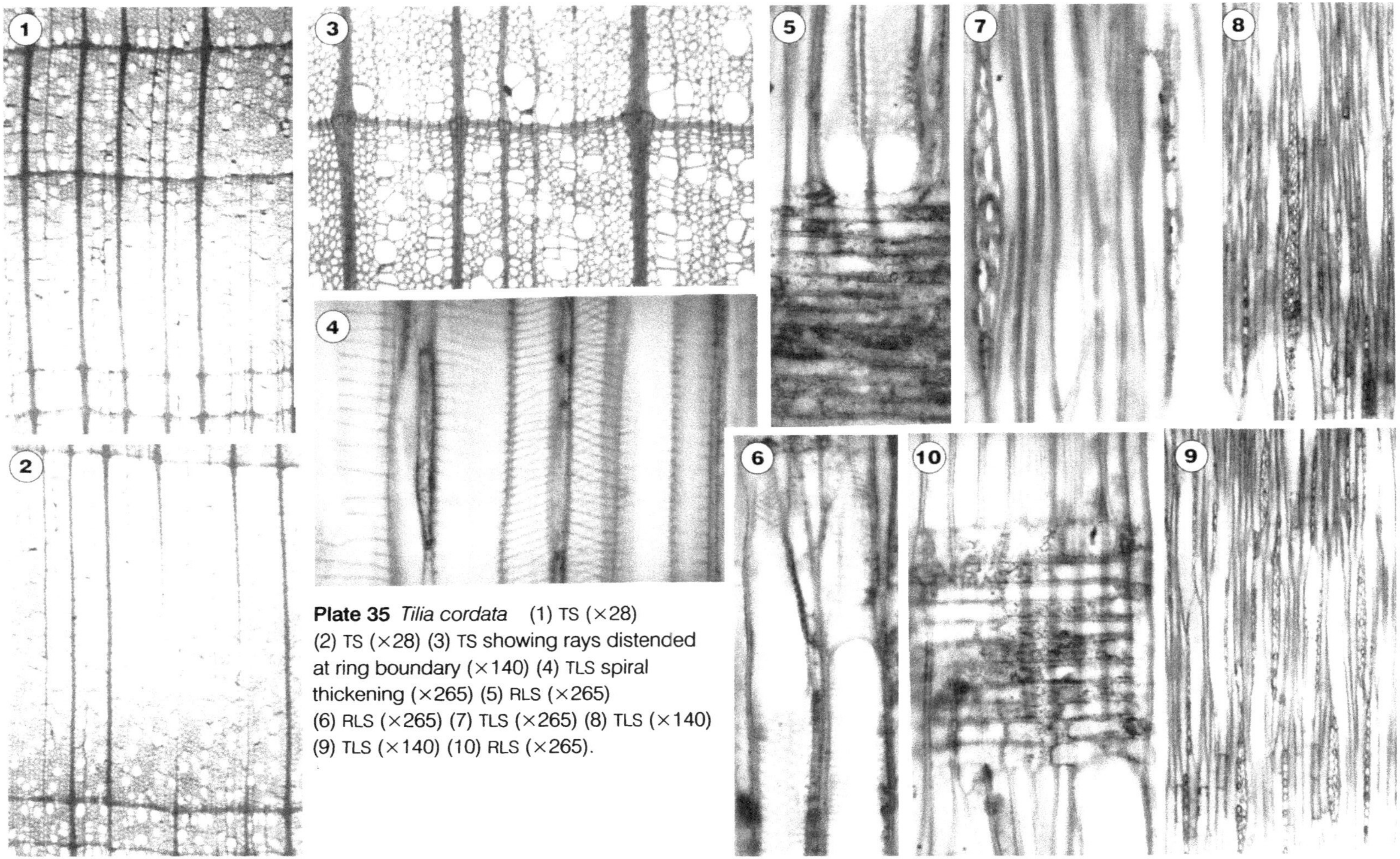

Plate 35 *Tilia cordata* (1) TS (×28) (2) TS (×28) (3) TS showing rays distended at ring boundary (×140) (4) TLS spiral thickening (×265) (5) RLS (×265) (6) RLS (×265) (7) TLS (×265) (8) TLS (×140) (9) TLS (×140) (10) RLS (×265).

Arbuteus unedo – Strawberry tree (Ericaceae)

Plate 36

Within northern Europe this small tree has a distribution limited to the southwest of Ireland but is also present in the west and central north Mediterranean region (see Map 104).

Arbuteus is a diffuse porous wood with solitary vessels as well as clusters and short radial files. Vessels are often a little angular. Ring boundaries are distinct due to numerous rows of terminal flattened cells. Unisereate rays are homogeneous, made up of upright cells only and up to 5 cells high though commonly only 2–3 cells high. Multisereate rays are heterogeneous with 2–3 rows of upright cells, occasionally more. The body of multisereate rays, made up of procumbent rays, tends to be rather short giving a spindle-shaped appearance to such rays. Spiral thickening is distinct. Simple and scalariform perforation plates are present though the latter with no more than 5 bars are present in smaller late wood vessels only.

The angularity of the vessels, the structure of the rays and presence of scalariform perforation plates is unlikely to allow confusion of *Arbuteus* with other woods.

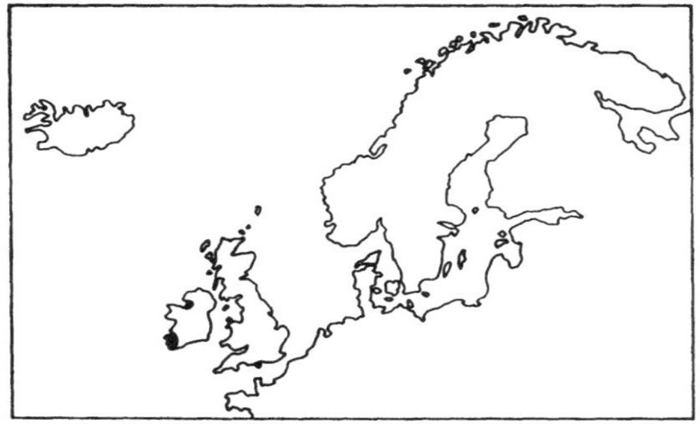

Map 104 Distribution of *Arbuteus unedo*

Table 2.36 Anatomical characters of *Arbuteus unedo* (Ericaceae)

Character	*Arbuteus unedo*	Character	*Arbuteus unedo*
TS		RLS	
Vessel arrangement	diffuse porous, semi-ring porous	Ray anatomy	heterogeneous types 1, 2, 3
Vessel grouping	solitary, short radial files, clusters	ray/vessel pits	enlarged
Tyloses	present	Libriform fibres	present (septate)
Parenchyma	apotracheal diffuse (paratracheal)	Fibre tracheids	present
Ground tissue	–	Vascular tracheids	absent
Ring boundary	distinct	Spiral thickening	present
Ray/ring boundary	–	Crystals	present
TLS		Perforation plates	simple (scalariform (1–4))
Ray width	1, 2–3(–4)		
Ray height	2–3, 5–15(–20)		
Ray cell shape	short fusiform		
Sheath cells	absent		

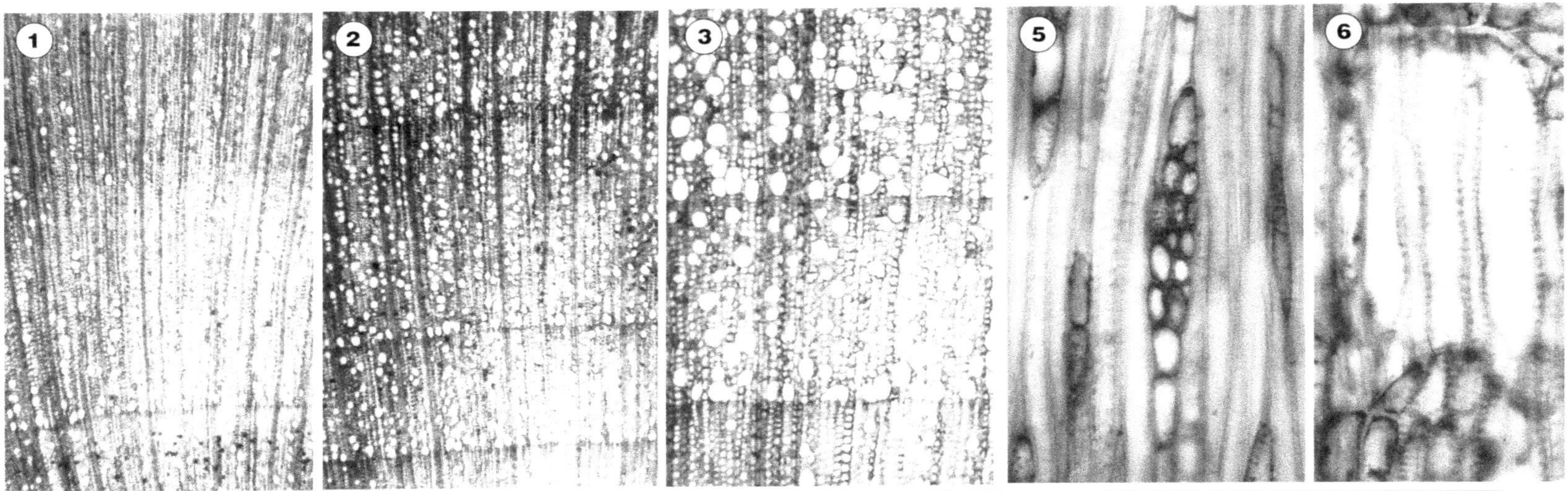

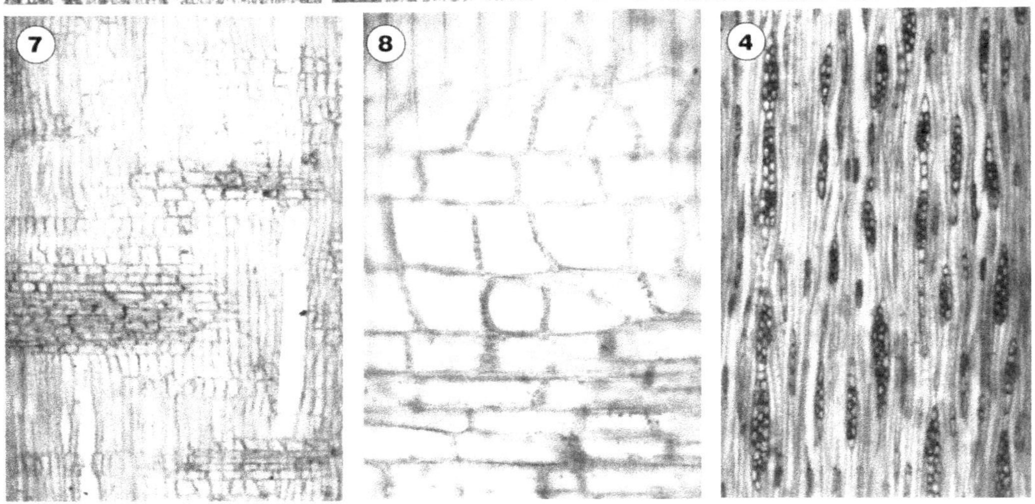

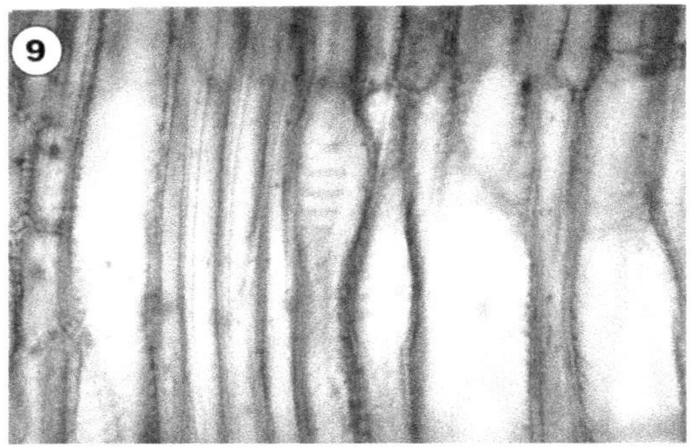

Plate 36 *Arbuteus unedo* (1) TS (×28) (2) TS (×28) (3) TS (×65) (4) TLS (×65) (5) TLS (×265) (6) RLS simple perforation plates (×265) (7) RLS rays (×65) (8) RLS ray (×265) (9) simple and scalariform perforation plates (×265)

Ficus carica – Fig (Moraceae)

Plate 37

The fig was introduced into northern Europe by Romans where it has been cultivated since and, in places, has naturalised. However, it does not compete well with native woodland trees and requires a very sheltered habitat to survive.

Ficus has diffuse porous wood with solitary vessels and short radial files of up to 3 and occasionally 4 vessels. Parenchyma is very noticeable in broad tangential bands against a thick-walled ground tissue. Growth boundaries are very difficult to determine. Rays are heterogeneous, commonly unisereate to 4-sereate, occasionally larger. Many unisereate rays are made up of upright cells only. The body cells of multisereate rays vary in size and rays tend to be somewhat irregular in shape. Perforation plates are simple.

The characteristic banded nature of the transverses section of *Ficus*, together with the distribution of the vessels and the structure of the rays is unlikely to cause confusion between this species and others.

Table 2.37 Anatomical characters of *Ficus carica* (Moraceae)

Character	*Ficus carica*	Character	*Ficus carica*
TS		RLS	
Vessel arrangement	diffuse porous	Ray anatomy	heterogeneous types 2, 3
Vessel grouping	solitary, short radial files	Libriform fibres	present
Tyloses	(present)	Fibre tracheids	absent
Parenchyma	banded	Vascular tracheids	absent
Ground tissue	thick-walled	Spiral thickening	absent
Ring boundary	very indistinct	Crystals	present
Ray/ring boundary	–	Perforation plates	simple
TLS			
Ray width	<3–4>		
Ray height	–30		
Sheath cells	absent		

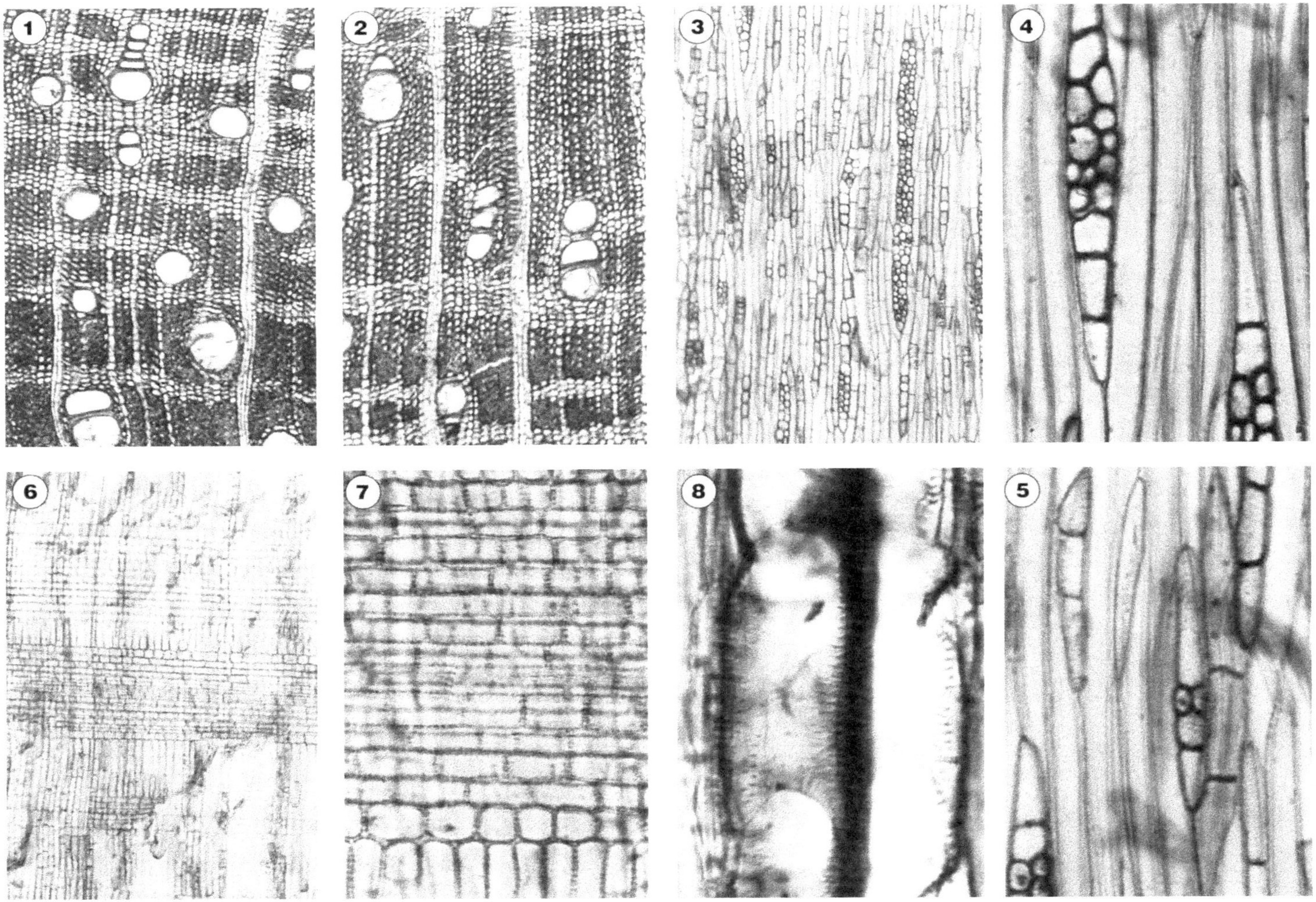

Plate 37 *Ficus carica* (1) TS (×65) (2) TS (×65) (3) TLS (×65) (4) TLS (×265) (5) TLS (×265) (6) RLS rays (×65) (7) RLS ray (×265) (8) RLS simple perforation plate (×265).

Prunus – Cherries (Prunoideae – Rosaceae)

Plate 38, Plate 39 and Plate 40

This is an important group of woodland trees and hedgerow and woodland margin shrubs distributed across much of northern Europe and Scandinavia. The group includes the following:

Taxon	Common name	Habit	Map
P. domestica (*P. insitia*)	Wild plum	small tree/large shrub	–
P. spinosa	Sloe/blackthorn	large thorny shrub	105
P. avium	Wild cherry	large tree	106
P. padus	Bird cherry	small tree	107

Anatomically all species are similar. *Prunus spinosa* and *P. domestica* cannot be separated. All three taxa are diffuse porous with a tendency towards the semi-ring porous state. Vessels are distributed in clusters of 3 to 4 vessels as well as being solitary. Ground tissue is thick-walled. Parenchyma is apotracheal diffuse but sparse. Unisereate rays are homogeneous with either procumbent or upright cells and up to 12 cells high, Multisereate rays up to 7 cells across (but see differentiating between the *Prunus* species below), up to 40 cells high, heterogeneous generally with one row of upright cells. Spiral thickening is present and perforation plates are simple.

Map 105 Distribution of *Prunus spinosa*

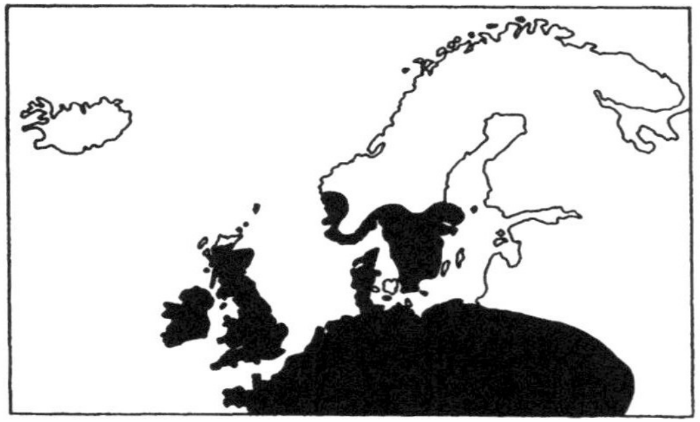

Map 106 Distribution of *Prunus avium*

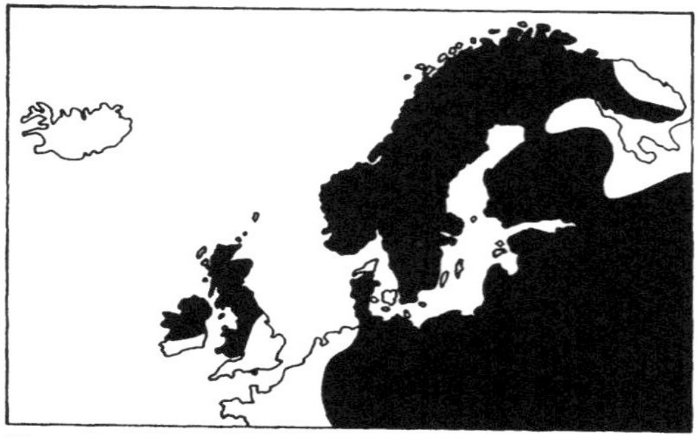

Map 107 Distribution of *Prunus padus*

Differentiating between the Prunus *species*

This is not easy and not always possible. Two characters are of potential diagnostic value. First is the presence/absence of libriform fibres and fibre tracheids.

Taxon	libriform fibres	fibre tracheids
P. spinosa P. domestica		present
P. avium	present	present
P. padus	present	

Libriform fibres and fibre tracheids are unfortunately very similar. They are both elongated fibres with tapering ends. Libriform fibres have simple pits which may be absent on the tangential wall whereas fibre tracheids have bordered pits often present on the tangential wall.

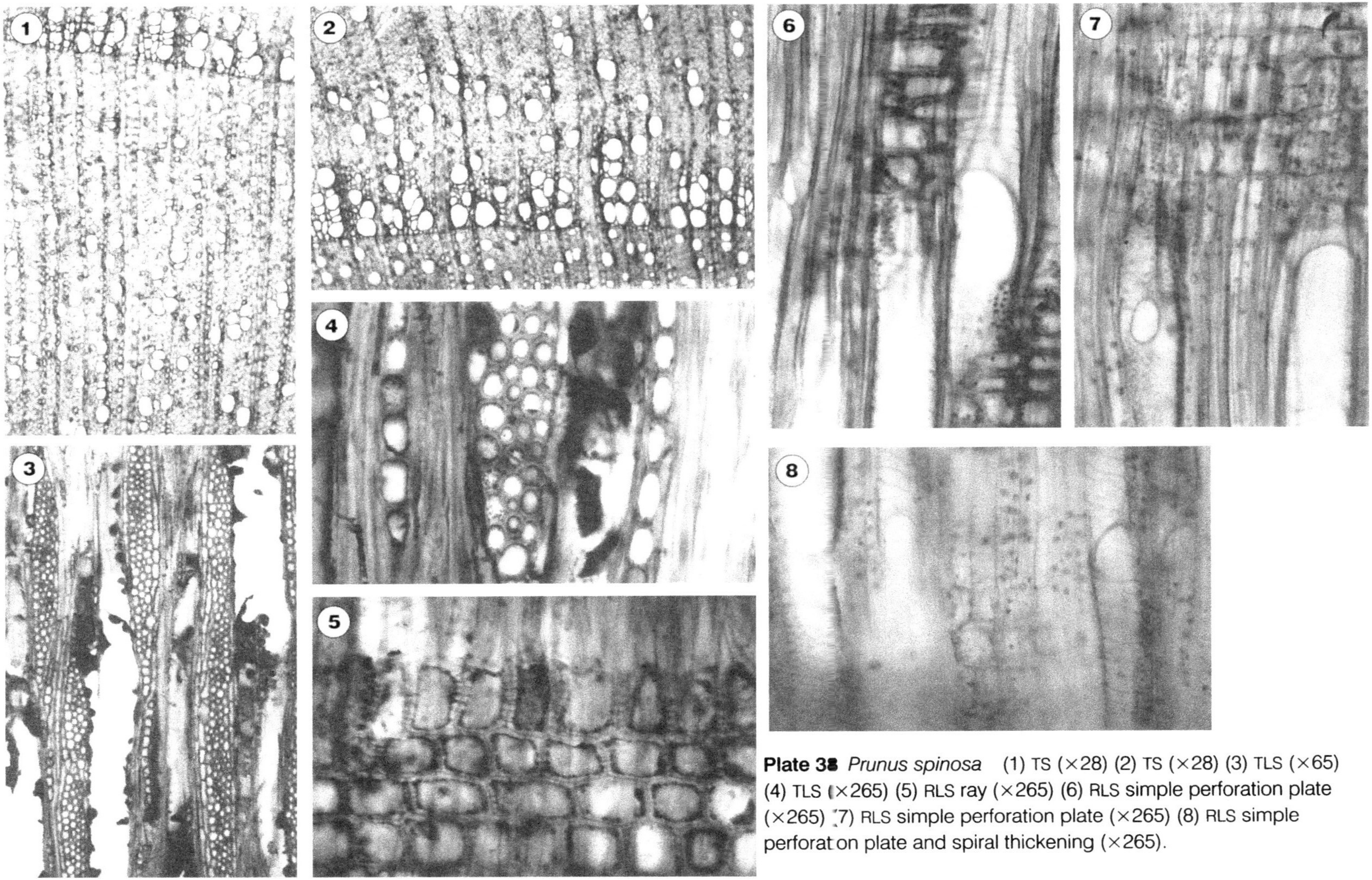

Plate 38 *Prunus spinosa* (1) TS (×28) (2) TS (×28) (3) TLS (×65) (4) TLS (×265) (5) RLS ray (×265) (6) RLS simple perforation plate (×265) (7) RLS simple perforation plate (×265) (8) RLS simple perforation plate and spiral thickening (×265).

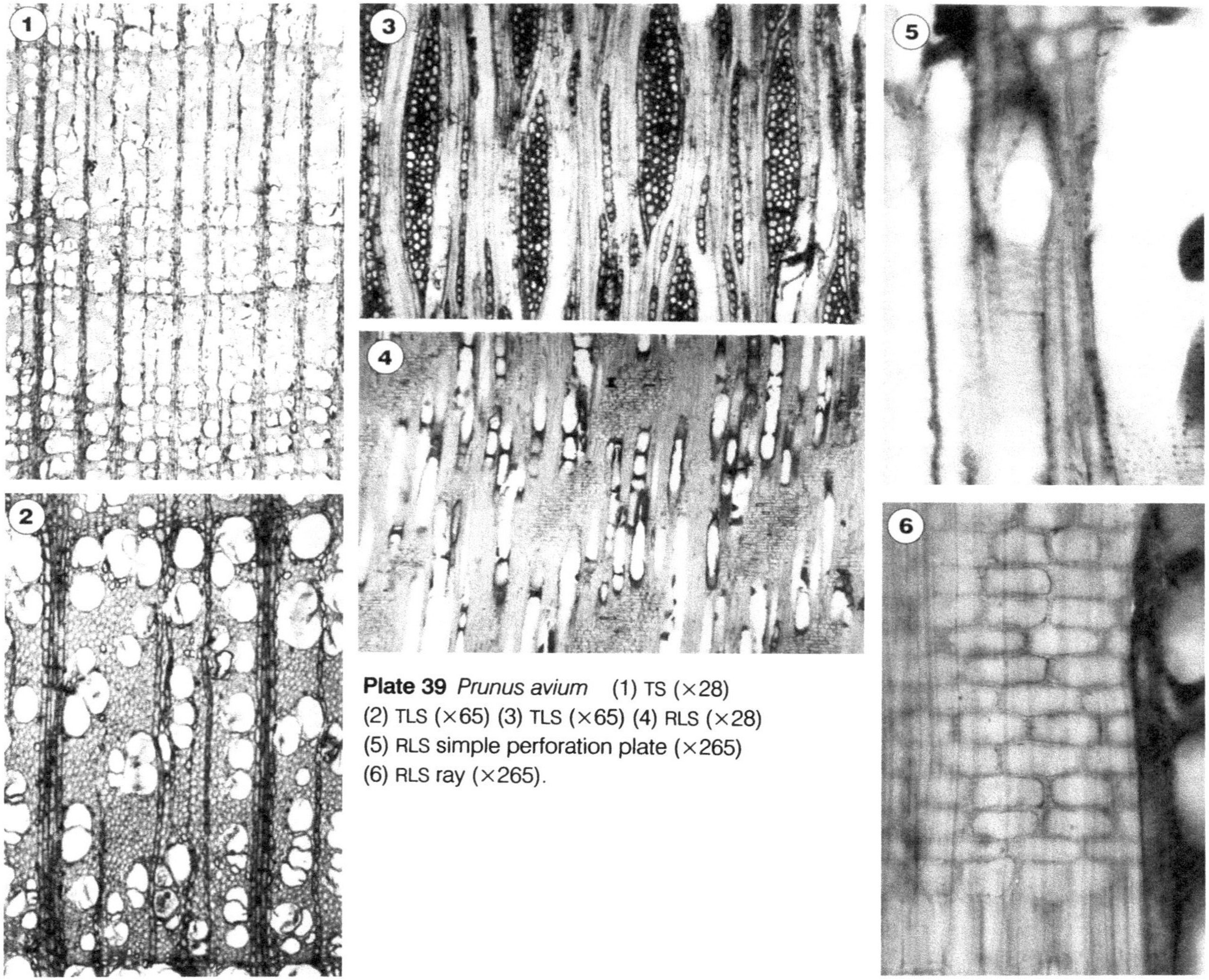

Plate 39 *Prunus avium* (1) TS (×28) (2) TLS (×65) (3) TLS (×65) (4) RLS (×28) (5) RLS simple perforation plate (×265) (6) RLS ray (×265).

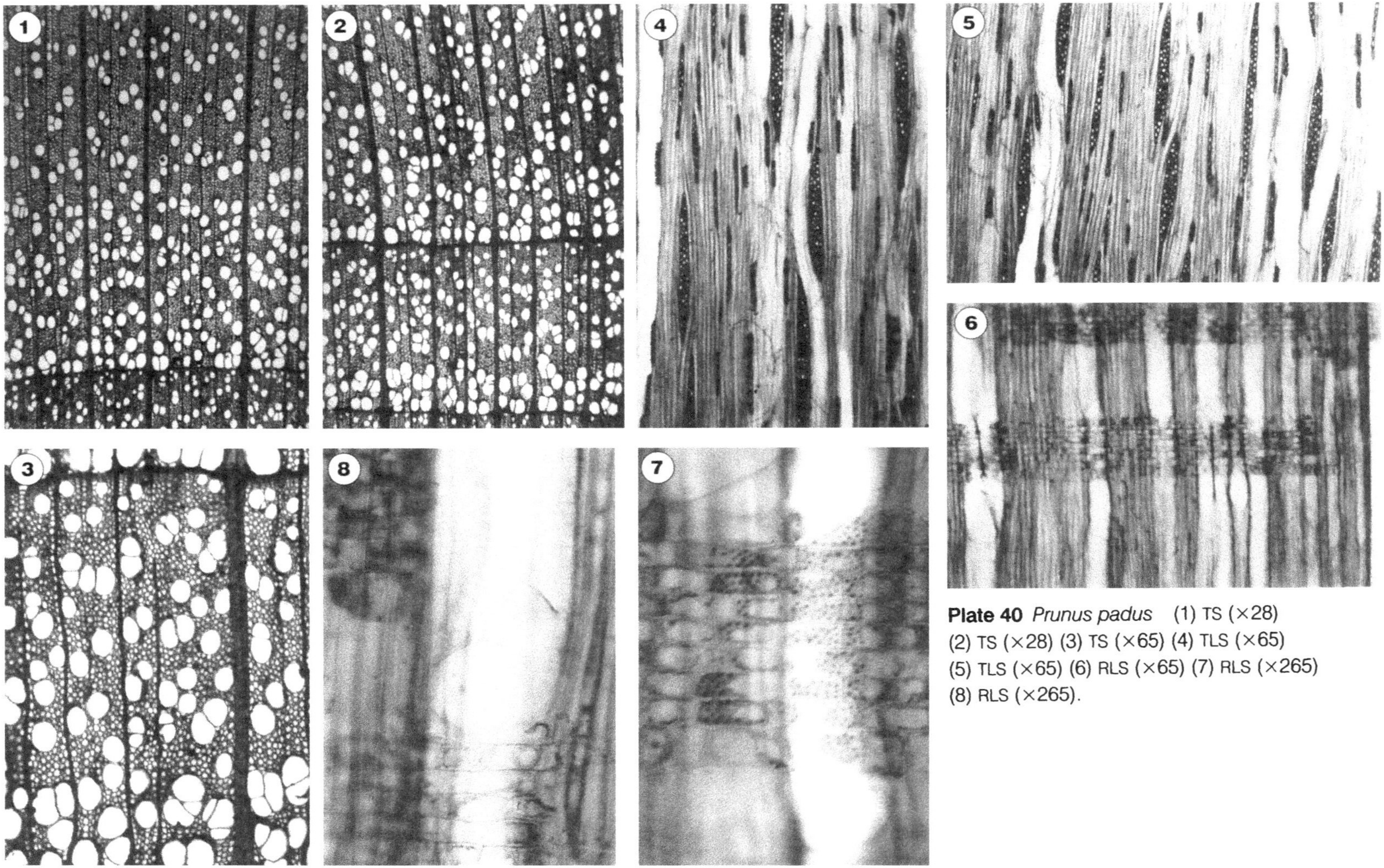

Plate 40 *Prunus padus* (1) TS (×28) (2) TS (×28) (3) TS (×65) (4) TLS (×65) (5) TLS (×65) (6) RLS (×65) (7) RLS (×265) (8) RLS (×265).

The second diagnostic character concerns the size of multisereate rays. *Prunus spinosa*/P. *domestica* has wide and tall rays, *P. padus* has wide but shorter rays and *P. avium* has narrower and less wide rays.

It must be stressed that with poor archaeological material or with limited number of rays to observe differentiation between the *Prunus* species is not always possible. Separation of *Prunus spinosa*/*P. domestica* from the other species based upon the ray height is probably the most secure use of these characters.

There is possible confusion with other diffuse porous woods with small multisereate rays – see *Frangula alnus*.

Taxon	Ray width	Ray height
P. spinosa P. domestica	3–7	50+
P. avium	2–4	up to 30
P. padus	2–7	up to 30

Table 2.38 Anatomical characters of the Prunoideae (Roscaeae)

Character	*Prunus domestica*/ *P. spinosa*	*Prunus avium*	*Prunus padus*
TS			
Vessel arrangement	diffuse porous	diffuse porous	diffuse porous
Vessel grouping	solitary, clusters	solitary, clusters	solitary, clusters
Tyloses	present	present	present
Parenchyma	apotracheal diffuse	apotracheal diffuse	apotracheal diffuse
Ground tissue	thick-walled	thick-walled	thick-walled
Ring boundary	distinct	distinct	distinct
Ray/ring boundary	–	–	–
TLS			
Ray width	1, 3–7	1, 2–4	1, 2–7
Ray height	–50+	–30	–30
Sheath cells	absent	absent	absent
RLS			
Ray anatomy	homogeneous, heterogeneous type 3	homogeneous, heterogeneous type 3	homogeneous, heterogeneous type 3
Libriform fibres	absent	present	present
Fibre tracheids	present	present	absent
Vascular tracheids	absent	absent	absent
Spiral thickening	present	present	present
Crystals	present	present	present
Perforation plates	simple	simple	simple

Acer – Maple species (Aceraceae)

Plate 41 and Plate 42

The five species in this group are all similar anatomically though they can be separated into two groups with difficulty (see table below). All are medium to large trees. *Acer pseudoplatanus* is a highly competitive weed in British woodlands, introduced in the 17th century.

Acer campestre – field maple (see Map 108)
Acer opalus – Italian maple (see Map 109)
Acer monspessulanum – Montpellier maple (see Map 110)
Acer platanoides – Norway maple (see Map 111)
Acer pseudoplatanus – sycamore (see Map 112)

Acer species are diffuse porous with solitary vessels and vessels in short radial files. Vessels are sometimes sparse and rarely crowded. Vessel size is mostly uniform across a year's growth. Ring boundaries are indistinct though terminal flattened cells are present. The fibre walls of the ground tissue vary in thickness resulting in a patchy appearance in TS. Rays are homogeneous up to 6 cells wide and up to 70 cells high (but see below – differentiating *Acer* sp.). Heterogeneous rays with one row of upright cells are rare. Spiral thickening is present and perforation plates are simple.

Differentiating the Acer sp.

It is possible to separate *Acer pseudoplatanus* and *A. platanoides* from *A. campestre, A. opalus* and *A. monspessulanum* on the basis of ray size. The *Acer campestre* group has rays that are generally 2–3 cells wide and no more than 30 cells high. *Acer pseudoplatanus* and *A. platanoides* have rays that are generally up to 7 cells across and up to 70 cells high. Though this seems to be a generally reliable character it is not always consistent and probably does not work for small diameter branches.

There is possible confusion with other diffuse porous woods with small multisereate rays – see *Frangula alnus*.

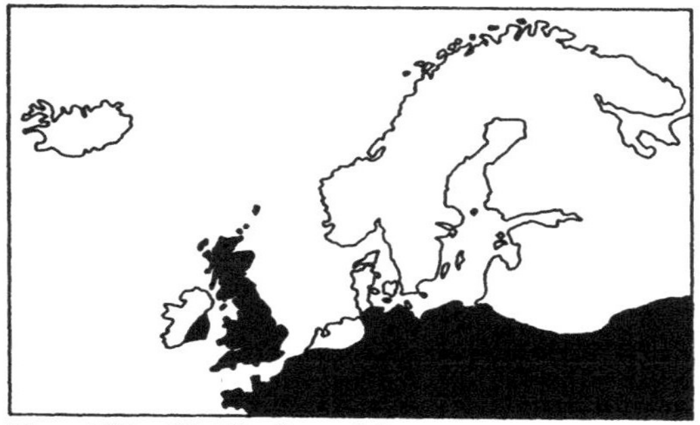

Map 108 Distribution of *Acer campestre*

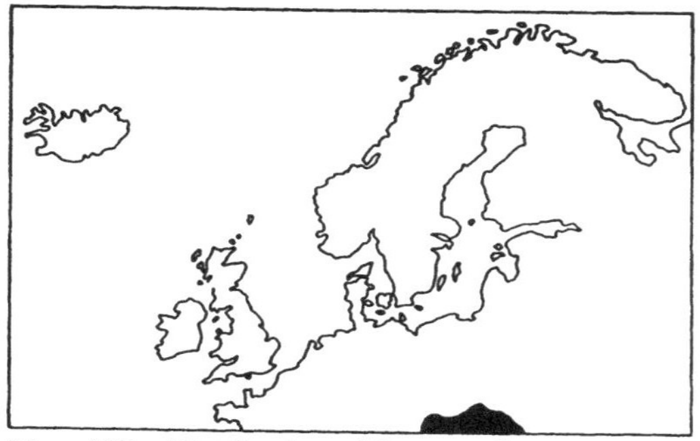

Map 109 Distribution of *Acer opalus*

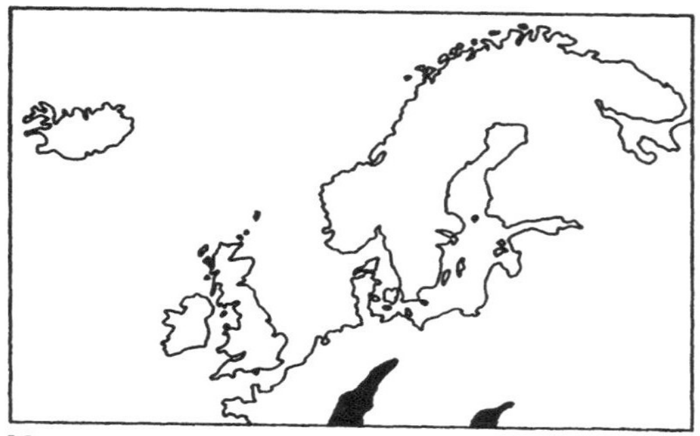

Map 110 Distribution of *Acer monspessulanum*

Table 2.39 Anatomical characters of the *Acer* (Aceraceae)

Character	*Acer campestre* *A. opalus* *A. monspessulanum*	*Acer platanoides* *A.pseudoplatanus*
TS		
Vessel arrangement	diffuse porous	diffuse porous
Vessel grouping	solitary, short radial files	solitary, short radial files
Tyloses	absent	absent
Parenchyma	apotracheal diffuse	apotracheal diffuse
Ground tissue	variable in thickness	variable in thickness
Ring boundary	indistinct	indistinct
Ray/ring boundary	(rays distended)	(rays distended)
TLS		
Ray width	(1) 2–3(–4)	(1,2) 3-6
Ray height	(5–) 20–30	(5–) 20–70
Sheath cells	absent	absent
RLS		
Ray anatomy	homogeneous	homogeneous
Libriform fibres	present	present
Fibre tracheids	absent	absent
Vascular tracheids	absent	absent
Spiral thickening	present	present
Crystals	absent	absent
Perforation plates	simple	simple

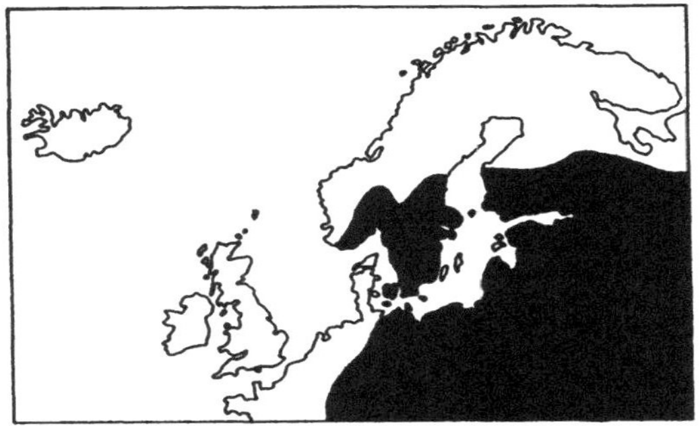

Map 111 Distribution of *Acer platanoides*

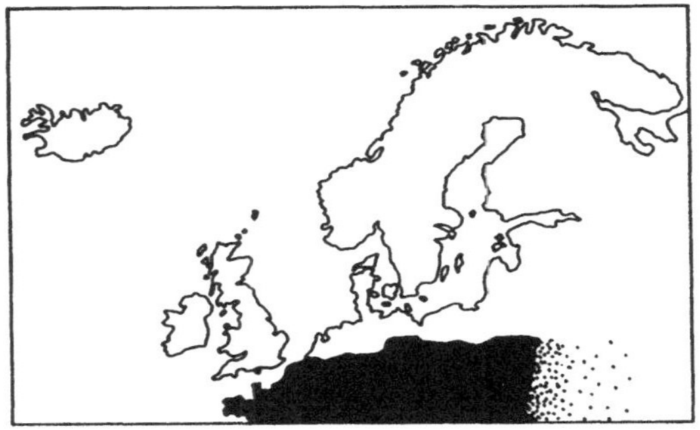

Map 112 Distribution of *Acer pseudoplatanus*

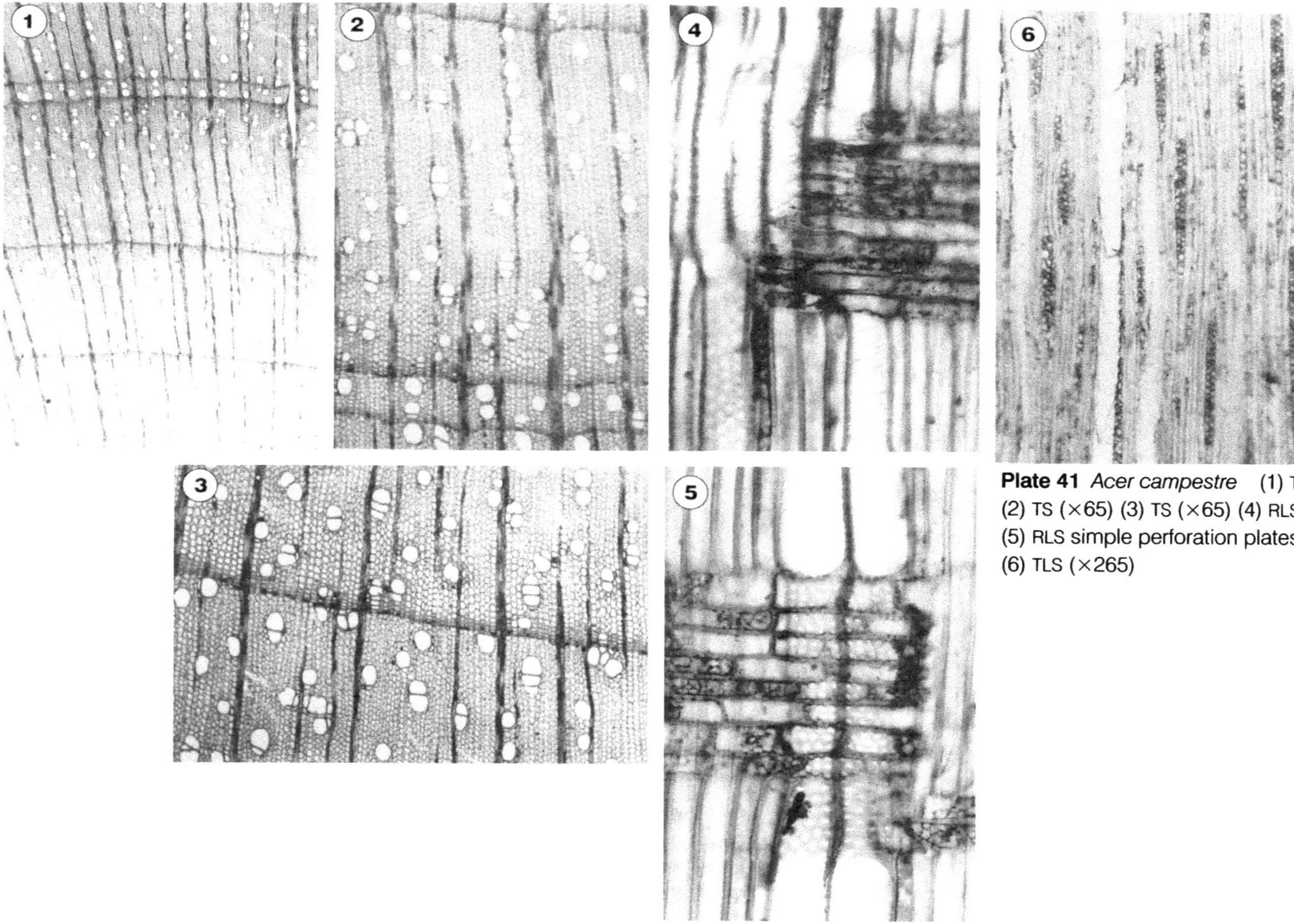

Plate 41 *Acer campestre* (1) TS (×28) (2) TS (×65) (3) TS (×65) (4) RLS (×265) (5) RLS simple perforation plates (×65) (6) TLS (×265)

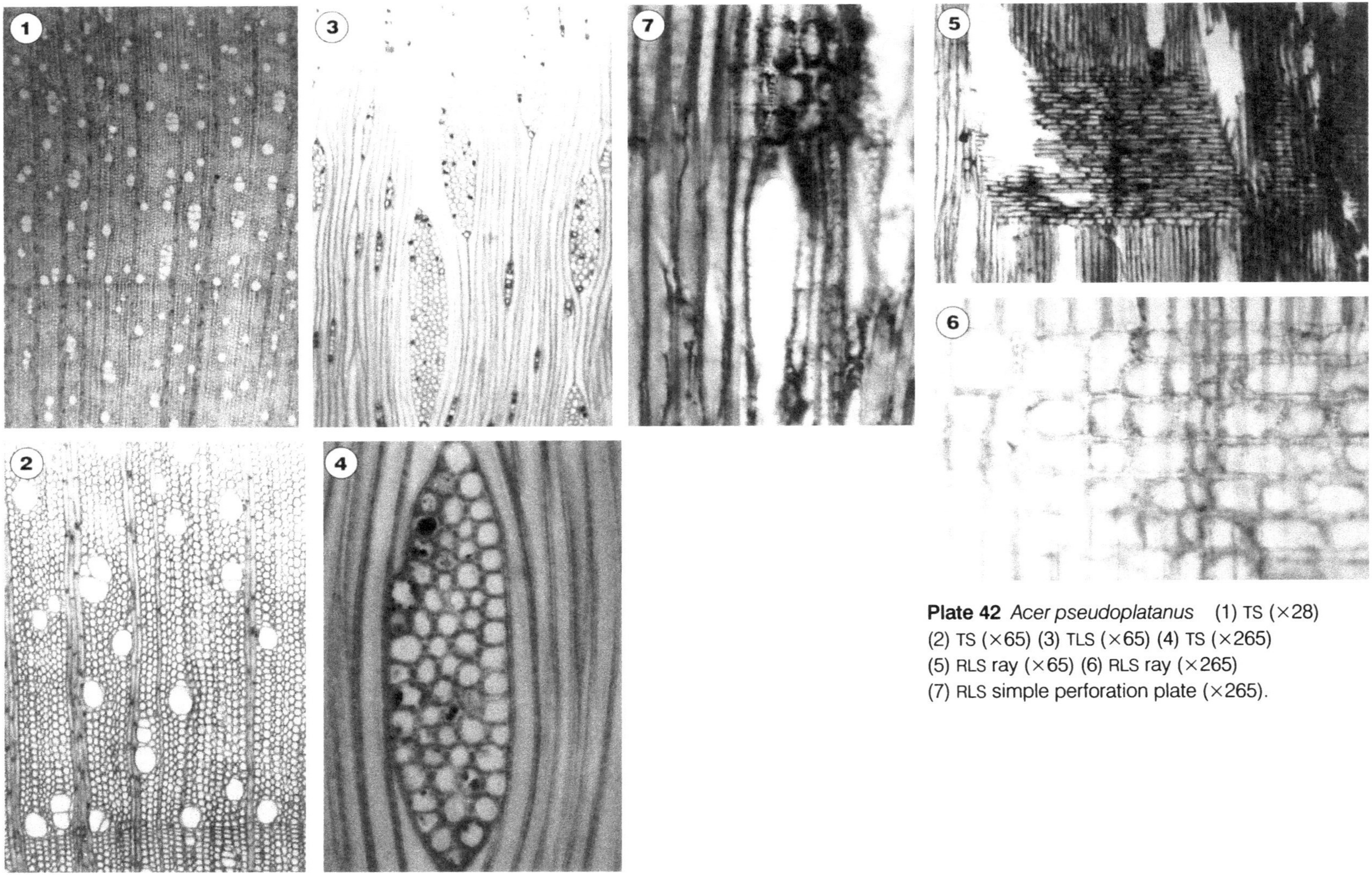

Plate 42 *Acer pseudoplatanus* (1) TS (×28) (2) TS (×65) (3) TLS (×65) (4) TS (×265) (5) RLS ray (×65) (6) RLS ray (×265) (7) RLS simple perforation plate (×265).

Juglans regia – Walnut (Juglandaceae)

Plate 43

Juglans regia is a large tree with a natural distribution in southeast Europe, though it was grown in northern Europe during the Roman period and has been planted and naturalised since. Distribution map 113 shows the modern naturalised distribution of *Juglans regia*.

Juglans is diffuse porous with large vessels distributed in short radial files as well as being solitary. Ring boundaries are distinct. Parenchyma is apotracheal sparse and in short tangential bands. Paratracheal parenchyma is present but rare. Rays are homogeneous and heterogeneous, mostly unisereate to 4-sereate sometimes larger. Perforation plates are simple.

There is possible confusion with other diffuse porous woods with small multisereate rays – see *Frangula alnus*.

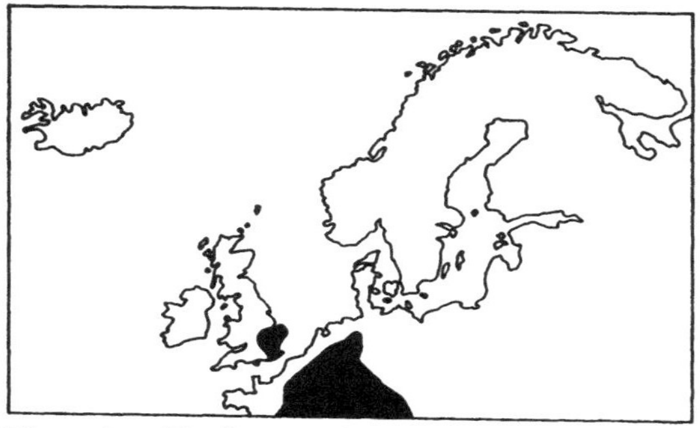

Map 113 Distribution of *Juglans regia*

Table 2.40 Anatomical characters of *Juglans regia* (Juglandaceae)

Character	*Juglans regia*	Character	*Juglans regia*
TS		RLS	
Vessel arrangement	diffuse porous to semi-ring porous	Ray anatomy	homogeneous (heterogeneous type 3)
Vessel grouping	solitary – short radial files	Libriform fibres	present
Tyloses	present	Fibre tracheids	absent
Parenchyma	apotracheal short tangential bands (paratracheal)	Vascular tracheids	absent
Ground tissue	–	Spiral thickening	absent
Ring boundary	–	Crystals	present
Ray/ring boundary	–	Perforation plates	simple
TLS			
Ray width	(1–) 2–4(–5)		
Ray height	15–30		
Sheath cells	absent		

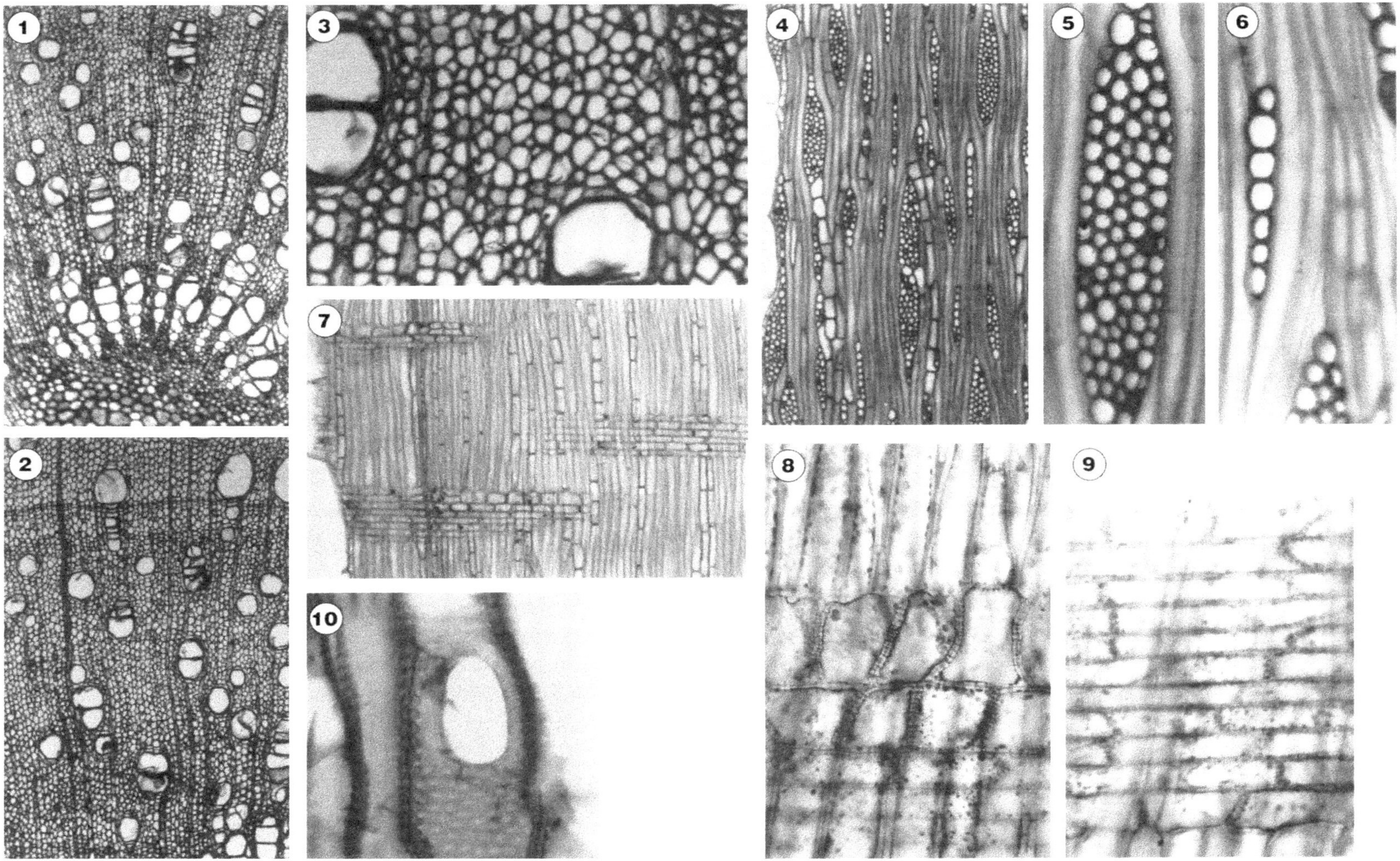

Plate 43 *Juglans regia* (1) TS pith and young wood (×28) (2) TS (×28) (3) TS (×265) (4) TLS (×65) (5) TLS (×265) (6) TLS (×265) (7) RLS (×65) (8) RLS ray (×265) (9) RLS ray (×265) (10) RLS simple perforation plate (×265).

Frangula alnus – Alder Buckthorn (Rhamnaceae)

Plate 44

Frangula is a small tree of woodland margins with a broad distribution over most of Europe and Scandinavia (see Map 114).

Frangula alnus is diffuse porous to semi-ring porous with vessels distributed in small clusters and short files as well as being solitary. Ground tissue is patchy due to varying thickness of fibre walls. Ring boundaries are generally distinct and rays are slightly distended at their intersection. Rays are heterogeneous with one row of upright cells, unisereate to trisereate, rarely larger, and up to 20 cells high though often shorter. Perforation plates are simple and spiral thickening is present, though fine.

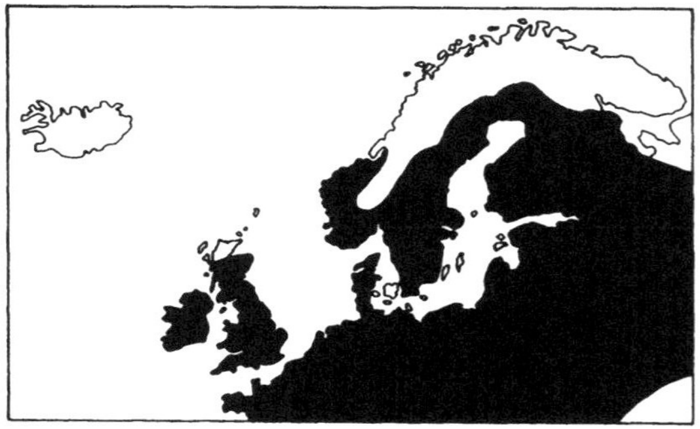

Map 114 Distribution of *Frangula alnus*

Table 2.41 Anatomical characters of *Frangula alnus* (Rhamnaceae)

Character	*Frangula alnus*
TS	
Vessel arrangement	semi-ring porous to diffuse porous
Vessel grouping	solitary, clusters and short radial files
Tyloses	absent
Parenchyma	paratracheal sparse
Ground tissue	–
Ring boundary	distinct
Ray/ring boundary	rays distended
TLS	
Ray width	1–3
Ray height	10–20
Sheath cells	absent
RLS	
Ray anatomy	heterogeneous type 3
Libriform fibres	present
Fibre tracheids	absent
Vascular tracheids	absent
Spiral thickening	fine
Crystals	absent
Perforation plates	simple

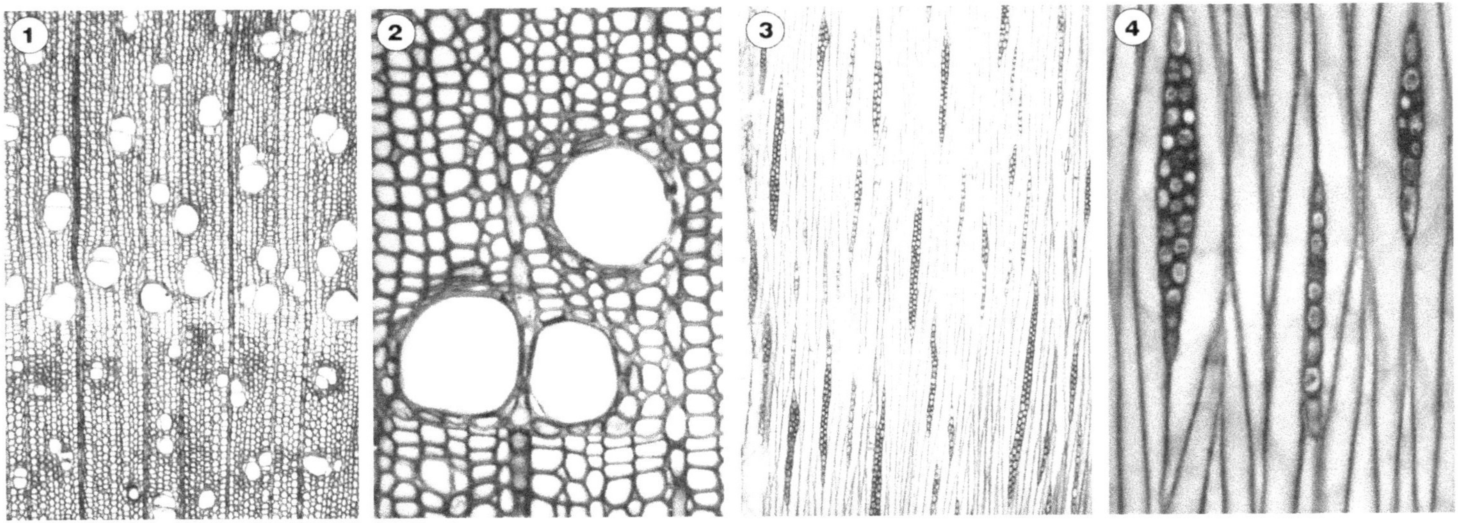

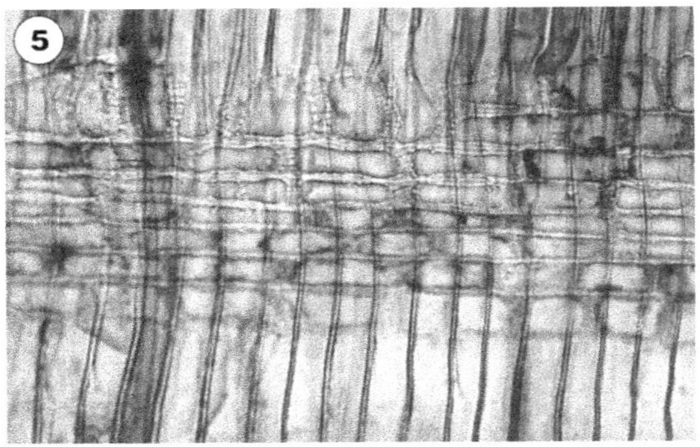

Plate 44 *Frangula alnus* (1) TS (×65) (2) TS (×265) (3) TLS (×65) (4) TLS (×265) (5) RLS ray (×265) (6) RLS simple perforation plate (×265)

There is possible confusion with other diffuse porous woods with small multisereate rays – *Prunus* spp., *Acer* spp. and *Juglans*. The following table may help.

Table 2.42 Diffuse porous woods summary table

Taxon	Vessel distribution	Vessel size	Ray width	Ray height	Ray structure	Spiral thickening
Prunus spinosa	solitary and in clusters	normal	1 3–7	–50+	homogeneous, heterogeneous type 3	present
Prunus avium	solitary and in clusters	normal	1, 2–4	–30	homogeneous, heterogeneous type 3	present
Prunus padus	solitary and in clusters	normal	1, 2–7	–30	homogeneous, heterogeneous type 3	present
Acer campestre	solitary and short radial files	normal	(1) 2–3(–4)	(5–) 20–30	homogeneous	present
Acer platanoides	solitary and short radial files	normal	(1,2) 3–6	(5–) 20–70	homogeneous	present
Juglans regia	solitary and short radial files	large	(1) 2–4(–5)	15–30	homogeneous (heterogeneous type 3)	absent
Frangula alnus	solitary, clusters and short radial files	normal	1–3	10–20	heterogeneous type 3	present (fine)

Group 6 – Diffuse porous woods with pores in long radial files

This relatively small group is defined by the presence of vessels in long radial files as well as solitary vessels and small clusters. All the taxa included are widely distributed. *Corylus avellana* and *Alnus glutinosa* are economically important as both are structural woods and for artefact manufacture.

General anatomy

All taxa share the characteristic of having long radial files of vessels, but the group is divided by the presence or absence of aggregate rays (see Table 2.43). The Betulaceae group with aggregate rays is not always easy to separate.

Table 2.43 Anatomical characters of diffuse porous woods with vessels in long radial files

Taxon	Aggregate rays present	Uniserate rays present	Largest ray width	Perforation plates	Spiral thickening present
Ilex aquifolium	no	yes	8	scalariform	yes
Alnus glutinosa/ incana	yes	yes	3	scalariform	no
Carpinus betulus	yes	yes	4	simple	no
Corylus avellana	yes	yes	3	scalariform	yes (fine)

Taxa descriptions

Group key

Aggregate rays absent	*Ilex*
Aggregate rays present	
• Sclariform perforation plates present	
– perforation plates with up to 10 bars	*Corylus*
– perforation plates with over 10 bars	*Alnus*
• Simple perforation plates present	*Carpinus*

Ilex aquifolium – Holly (Aquifoliaceae)

Plate 45

Ilex aquifolium is a medium-sized tree of deciduous woodland with a distribution throughout western Europe and the western fringes of Scandinavia (see Map 115).

Ilex aquifolium is diffuse porous with vessels distributed in long radial files as well as few solitary vessels and clusters. Radial files can have up to 12 vessels in them. Parenchyma is paratracheal and apotracheal. The ring boundary is marked by up to four rows of slightly flattened cells at the end of a year's growth. Rays are heterogeneous with many rows of upright cells. Unisereate rays are composed of upright cells only and up to 8 cells high. Multisereate rays are 4 to 8 cells across and up to 3 mm high. Perforation plates are tall and scalariform with up to 30 bars. Spiral thickening is distinct.

Only the Betulaceae group described below (not including betula itself) has similarly long radial files and none of these species has large rays, but all of them possess aggregate rays (see below).

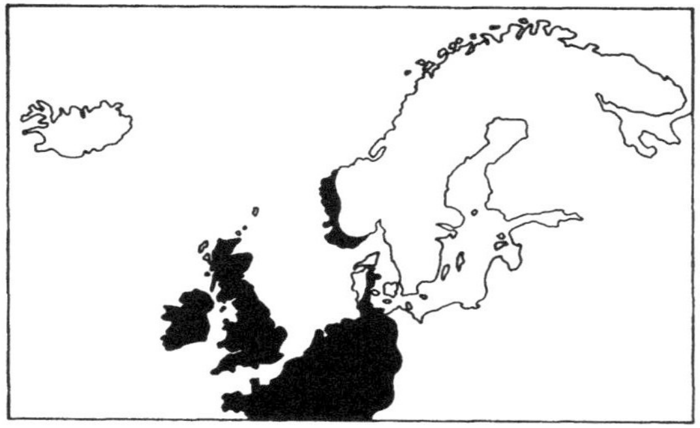

Map 115 Distribution of *Ilex aquifolium*

Table 2.44 Anatomical characters of *Ilex aquifolium* (Aquifoliaceae)

Character	*Ilex aquifolium*	Character	*Ilex aquifolium*
TS		RLS	
Vessel arrangement	diffuse porous	Ray anatomy	heterogeneous type 1
Vessel grouping	long radial files	Libriform fibres	absent
Tyloses	absent	Fibre tracheids	present
Parenchyma	apotracheal diffuse, paratracheal	Vascular tracheids	absent
Ground tissue	uniform	Spiral thickening	present
Ring boundary	distinct	Perforation plates	scalariform, 10–30
Ray/ring boundary	–		
TLS			
Ray width	1, 4–8		
Ray height	2–8, 3 mm		
Sheath cells	absent		

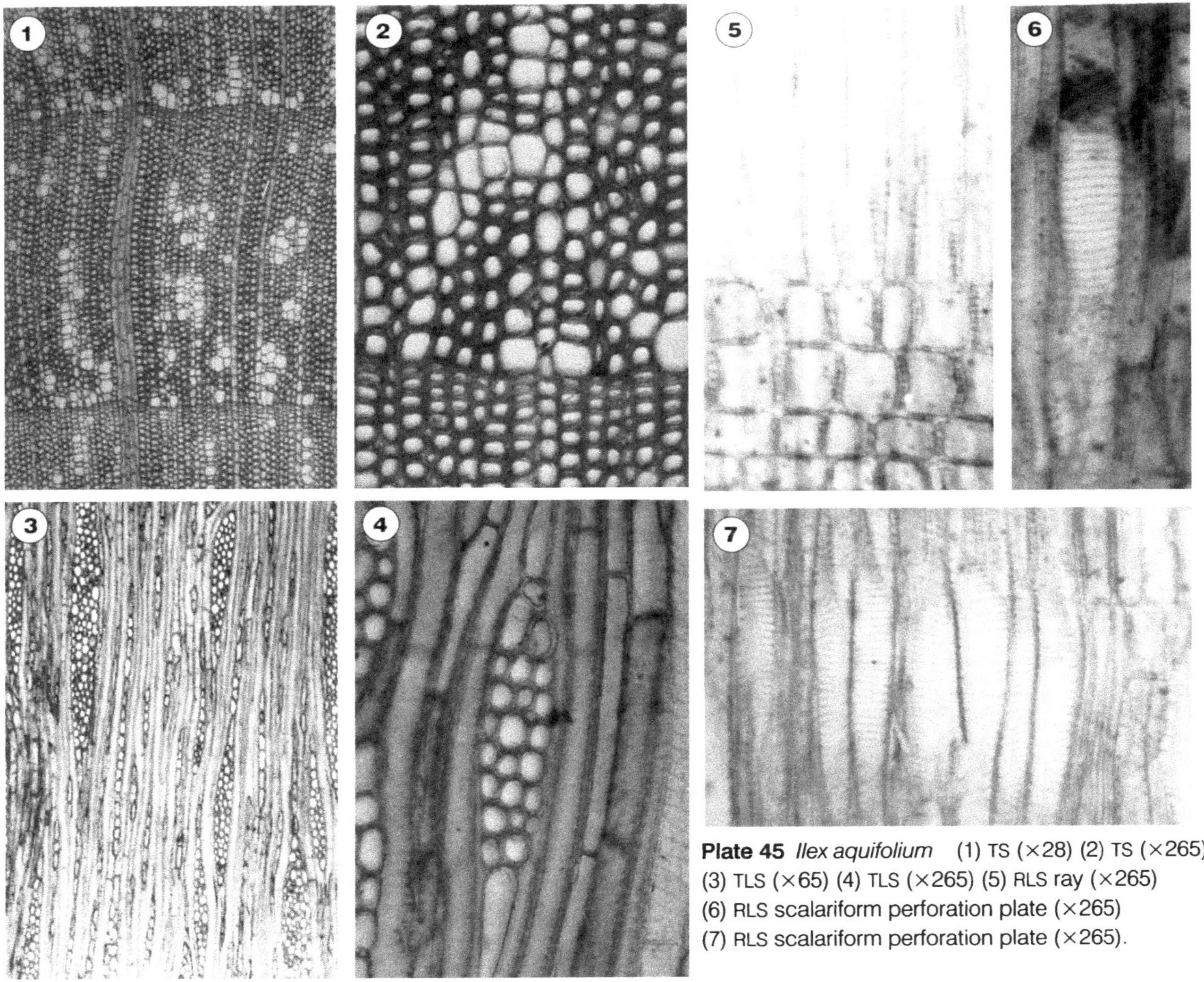

Plate 45 *Ilex aquifolium* (1) TS (×28) (2) TS (×265) (3) TLS (×65) (4) TLS (×265) (5) RLS ray (×265) (6) RLS scalariform perforation plate (×265) (7) RLS scalariform perforation plate (×265).

Corylus avellana – Hazel

Alnus sp. – Alder

Carpinus betulus – Hornbeam (Betulaceae)

Plate 46, Plate 47 and Plate 48

These three species all have aggregate rays and have other features in common. *Corylus avellana* is commonly seen as a managed medium to large shrub with a distribution across western northern Europe and north into Scandinavia (see Map 116). There are two *Alnus* species with a natural distribution in northern Europe, *A. glutinosa* with a western European and Baltic distribution, and *A. incana* which has a more central and northerly distribution reaching into the far north of Scandinavia (see Maps 117 and 118). Both are large economically important trees. *Carpinus betulus* is also a medium to large tree with a western European distribution reaching into the southern tip of Scandinavia (see Map 119).

All three taxa (the two *Alnus* species cannot be differentiated anatomically) possess aggregate rays. These are aggregates of unisereate and small multisereate rays that appear as a large multisereate ray. In all three species they are frequent and an important diagnostic character. Vessels in the three taxa are distributed in long radial files of up to 12 vessels. Parenchyma is apotracheal diffuse and in clusters in *Corylus*. The ring boundary in *Corylus* is strongly undulating; in *Alnus* it undulates markedly at the position of the aggregate rays, but weakly elsewhere; in *Carpinus* it is weakly undulating. *Corylus* alone has fine spiral thickening.

The three taxa can be differentiated on the basis of ray anatomy and the perforation plates.

- Rays in *Alnus* are homogeneous and unisereate up to 25 cells high; unisereate to trisereate rays occur within the aggregate rays. Perforation plates are scalariform with up to 25 fine bars.
- Rays in *Corylus* are mostly heterogeneous with some homogeneous rays, mostly unisereate, with some bisereate rays and up to 40 cells high; unisereate to trisereate rays occur within the aggregate rays. Perforation plates are scalariform with up to 10 distinct bars.

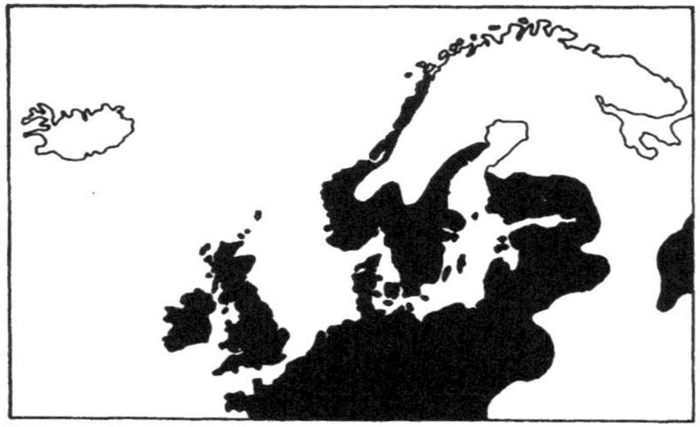

Map 116 Distribution of *Corylus avellana*

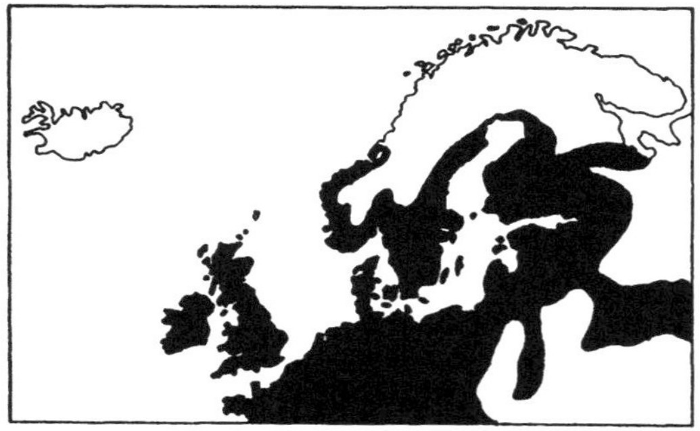

Map 117 Distribution of *Alnus glutinosa*

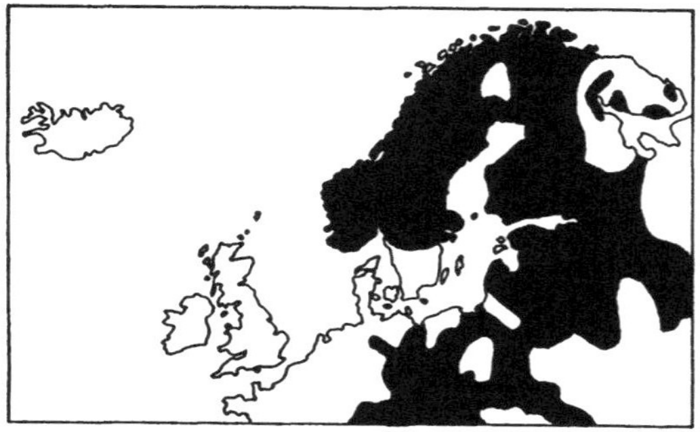

Map 118 Distribution of *Alnus incana*

Table 2.45 Summary of the main characters used to differentiate taxa with aggregate rays

Taxa/character state	Ring boundary	Ray width	Ray width (at AR)	Ray height	Ray anatomy	Perforation plates
Alnus sp	undulating (AR)	1	1–3	3–25	Hm	Scalariform 10–20
Corylus avellana	undulating	1(–2)	1–3	10–25(–40)	Ht 2–3 (Hm)	Scalariform 5–10
Carpinus betulus	(undulating)	1–2	1–4	10–20+	Ht 2–3 (Hm)	Simple

- Rays in *Carpinus* are mostly heterogeneous with some homogeneous rays, unisereate and bisereate and mostly up to 20 cells high; unisereate to 4-sereate rays occur in the aggregate rays. Perforation plates are simple often with rather straight sides.

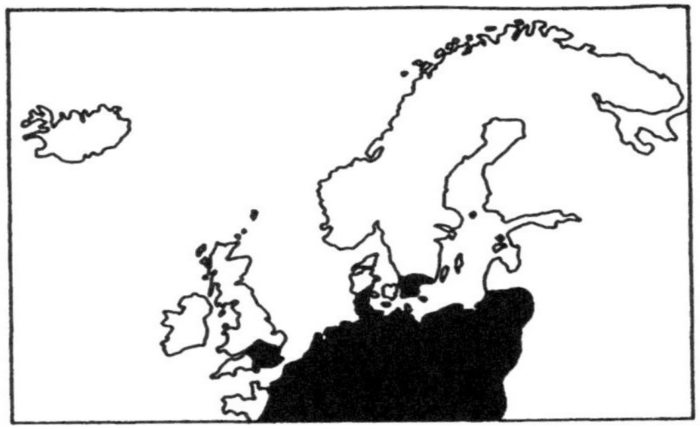

Map 119 Distribution of *Carpinus betulus*

There is a degree of similarity between the three taxa, perhaps especially between *Alnus* and *Corylus*, and in poorly preserved wood or where sampling is restricted, perforation plates may not be visible and a secure identification may not be possible.

The presence of aggregate rays, long radial files and unisereate and small multisereate rays easily differentiates this group from most other woods. *Betula pendula* and *B. pubescens*, however, can have aggregate rays, though these are rare and were not found in the material examined. Neither *Betula* species has long radial files, though they have perforation plates very similar to those of *Alnus*. Betula rays are up to 4-sereate.

Table 2.46 Anatomical characters of Betulaceae taxa with aggregate rays

Character	*Corylus avellana*	*Alnus*[1]	*Carpinus betulus*
TS			
Vessel arrangement	diffuse porous, semi-ring porous	diffuse porous, semi-ring porous	diffuse porous, semi-ring porous
Vessel grouping	long radial files, solitary, clusters	long radial files, solitary, clusters	long radial files, solitary, clusters
Tyloses	–	–	–
Parenchyma	apotracheal (diffuse and in clusters)	apotracheal diffuse	apotracheal diffuse
Ground tissue	–	–	–
Ring boundary	undulating	undulating (at aggregate ray)	(undulating)
Ray/ring boundary	distinct	distinct	indistinct
Aggregate rays	present	present	present
TLS			
Ray width	1(–2)	1	1–2
Ray width (aggregate rays)	1–3	1–3	1–4
Ray height	10–25(–40)	3–25	10–20+
Sheath cells	absent	absent	absent
RLS			
Ray anatomy	heterogeneous types 2–3 (homogeneous)	homogeneous	heterogeneous types 2–3 (homogeneous)
Libriform fibres	present	present	present
Fibre tracheids	present	(present)	(present)
Vascular tracheids	absent	absent	absent
Spiral thickening	present (fine)	absent	absent
Perforation plates	scalariform 5–10	scalariform 10–20+	simple

1. *Alnus* includes *A. glutinosa* and *A. incana*

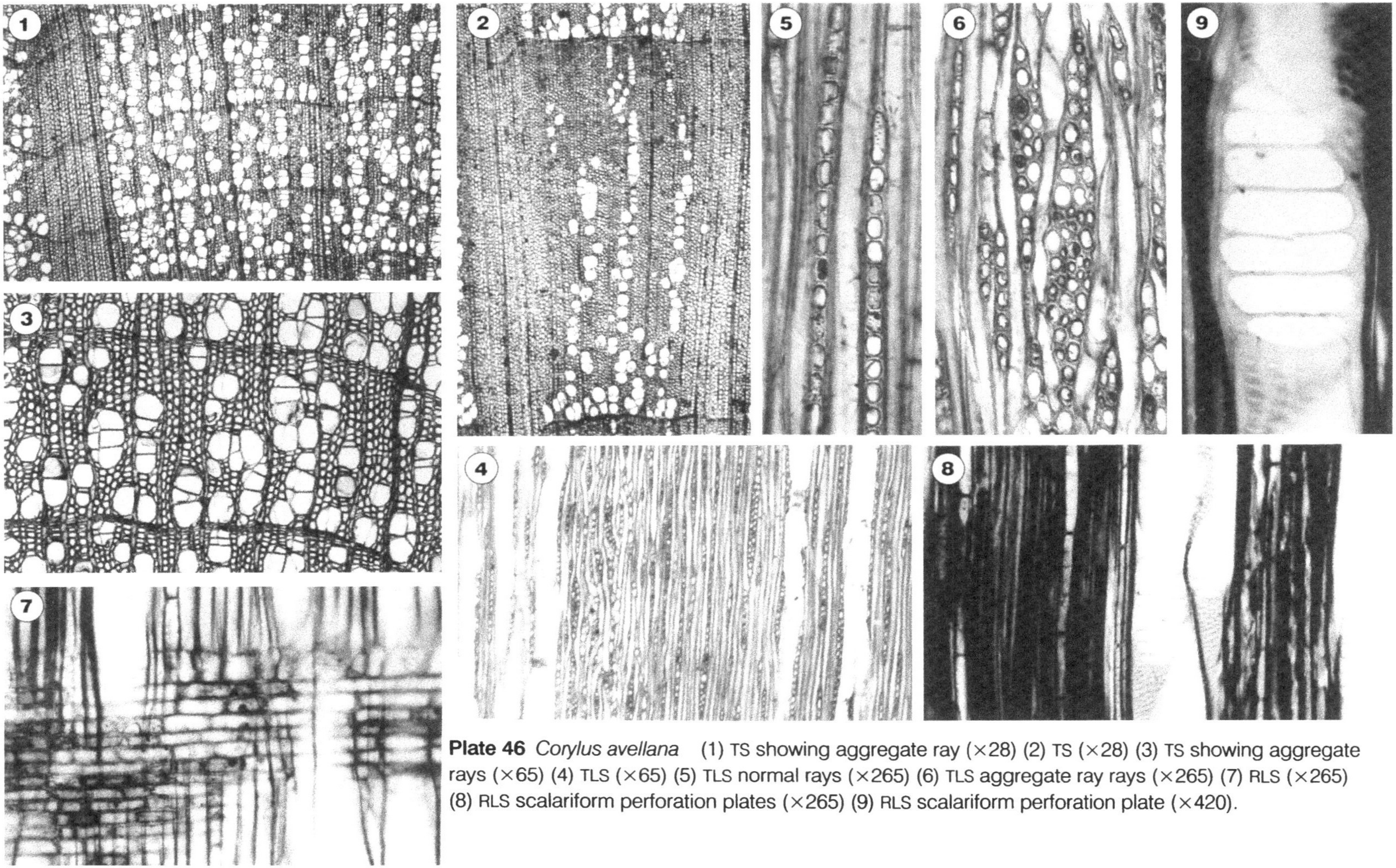

Plate 46 *Corylus avellana* (1) TS showing aggregate ray (×28) (2) TS (×28) (3) TS showing aggregate rays (×65) (4) TLS (×65) (5) TLS normal rays (×265) (6) TLS aggregate ray rays (×265) (7) RLS (×265) (8) RLS scalariform perforation plates (×265) (9) RLS scalariform perforation plate (×420).

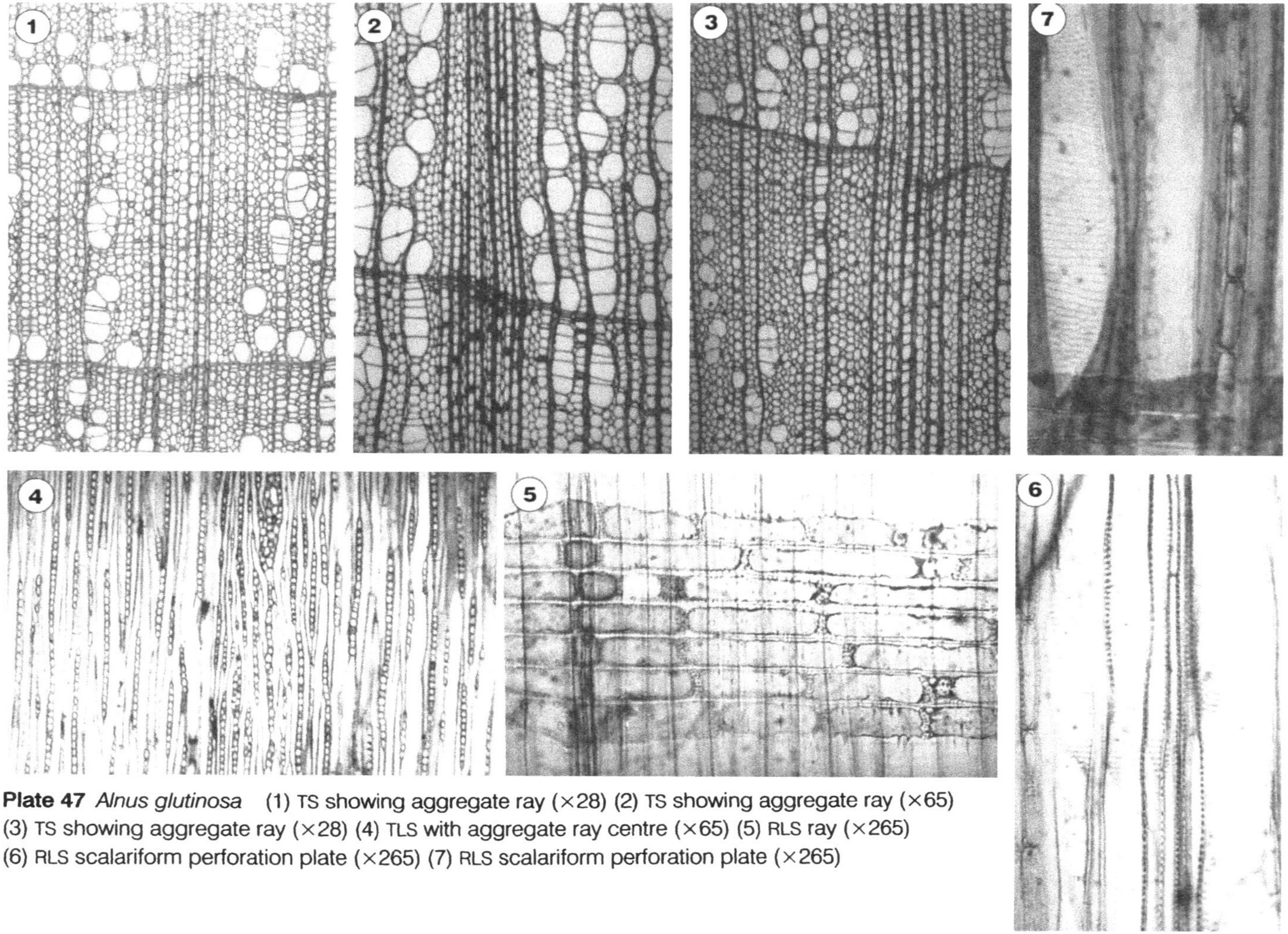

Plate 47 *Alnus glutinosa* (1) TS showing aggregate ray (×28) (2) TS showing aggregate ray (×65) (3) TS showing aggregate ray (×28) (4) TLS with aggregate ray centre (×65) (5) RLS ray (×265) (6) RLS scalariform perforation plate (×265) (7) RLS scalariform perforation plate (×265)

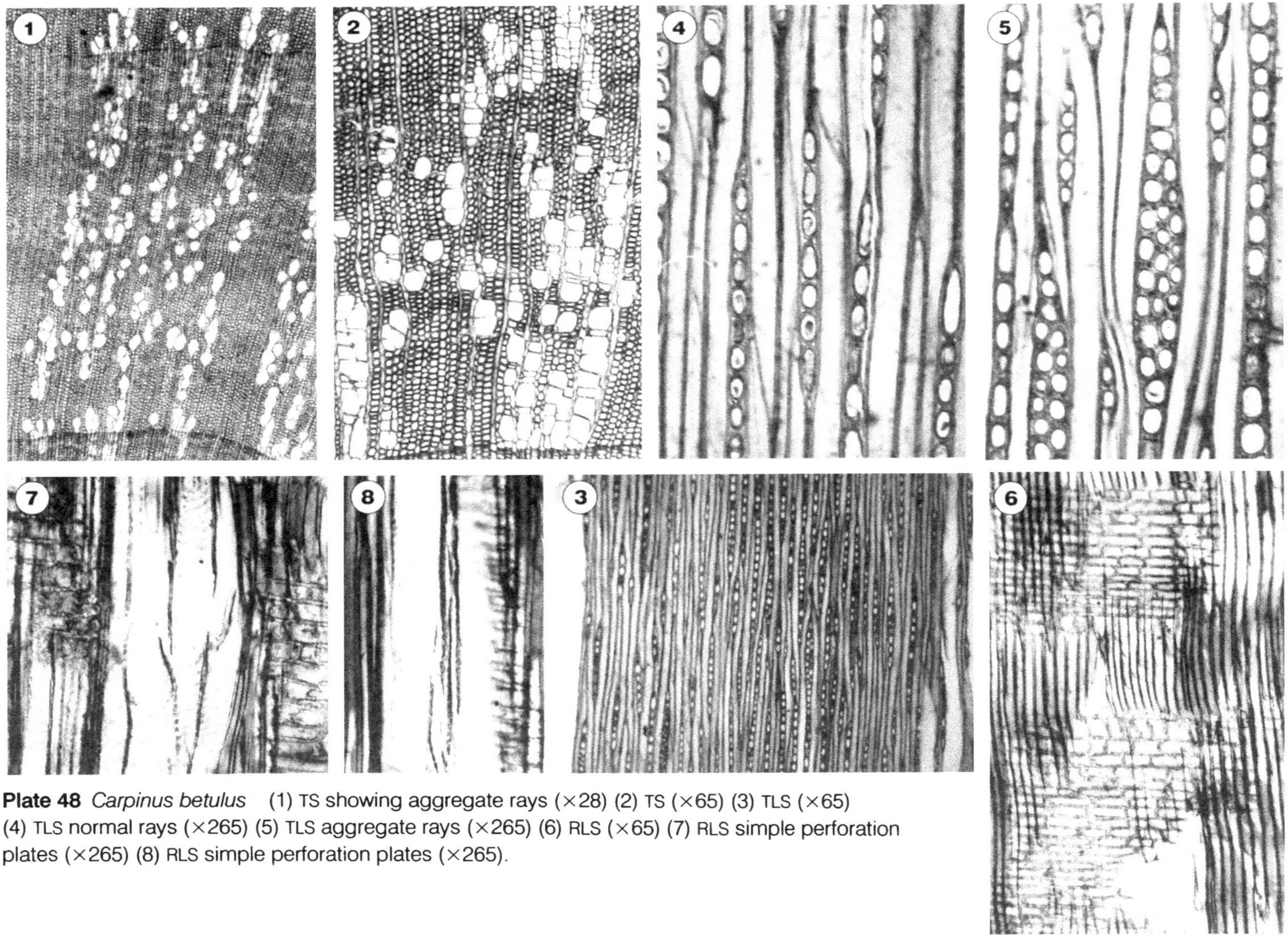

Plate 48 *Carpinus betulus* (1) TS showing aggregate rays (×28) (2) TS (×65) (3) TLS (×65) (4) TLS normal rays (×265) (5) TLS aggregate rays (×265) (6) RLS (×65) (7) RLS simple perforation plates (×265) (8) RLS simple perforation plates (×265).

Group 7 – Diffuse porous woods with pores in clusters

There are only two taxa with vessels in pore clusters to the exclusion of vessels in files or solitary vessels. The late wood of *Ulmus* also comes under this heading and is dealt with below. *Hedera helix* is a vine of woodland margins and hedgerows. Mature plants can be well over 20 cm in diameter at their base though wood of this size is limited even on very large plants. *Sambucus nigra* is a large shrub and *Sambucus racemosa* is a small tree. Both are plants of woodland margins and hedgerows.

General anatomy

Table 2.47 Summary of the anatomical characters of the diffuse porous woods with pores in clusters

Taxon	Vessel grouping	Uniserеate rays present	Largest ray width	Perforation plates	Spiral thickening present
Sambucus nigra *Sambucus racemosa*	clusters	no	4	simple	no
Hedera helix	tangential clusters	no	20	simple	no

The late wood of *Ulmus* species also has vessels in tangential clusters and may be confused with *Hedera helix* (see below) though ray height in the latter is far greater and *Ulmus* has spiral thickening present.

Taxa descriptions

Group key

Vessel clusters irregular, rays up to 4 cells across — *Sambucus*

Vessel clusters distinctly tangential, rays up to 20 cells across — *Hedera*

Sambucus nigra – Elder

Sambucus racemosa – Red-berried elder (Caprifoliaceae)

Plate 49 and Plate 50

Sambucus nigra is a large shrub of hedgerows across much of northern Europe reaching into southern Scandinavia (see Map 120). *S. racemosa* has a more limited westerly distribution but reaches further north into Sweden and southern Finland (see Map 121).

Both these species have a semi-ring porous to diffuse porous vessel arrangement with vessels in small irregular clusters with radial, tangential and oblique orientations. There are very few solitary vessels. Ground tissue is generally thick-walled and parenchyma is apotracheal. Rays are 2 to 4 cells wide and between 10 and 30 cells high, heterogeneous and with sheath cells. Perforation plates are simple.

According to Schweingruber (1990) differentiating between *Sambucus nigra* and *Sambucus racemosa* is based on three characters. First, the ring boundary in *S. nigra* has about 3 rows of flattened cells (vascular tracheids) while *S. racemosa* has about 10. The second character concerns the ray cell walls which are thick-walled and beaded in *S. nigra* and thin-walled and not beaded in *S. racemosa*. A final character concerns the ray height which is on average about 20 in *S. nigra* and about 15 in *S. racemosa*.

In the material examined here the number of flattened cells at the end of a year's growth seen in TS is similar in both species, although more continuous in *S. racemosa* than in *S. nigra*. In both species the ray walls were beaded and similarly thick. Ray height differs and seems to be the most useful of these characters.

The anatomy of the two *Sambucus* species would be difficult to confuse with other species.

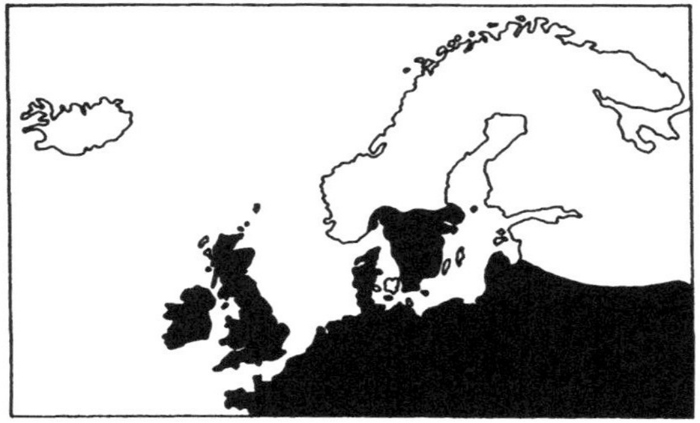

Map 120 Distribution of *Sambucus nigra*

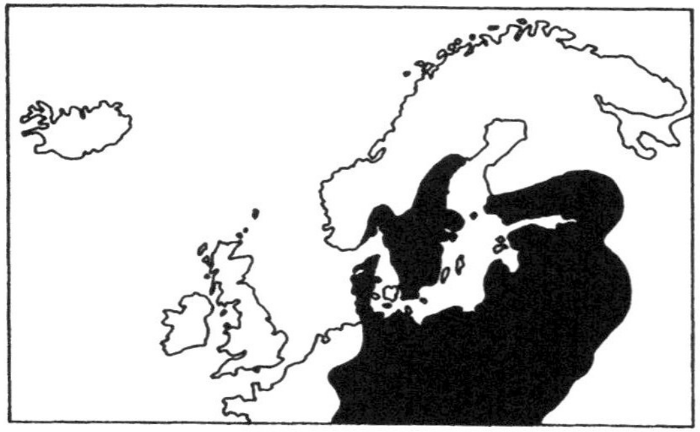

Map 121 Distribution of *Sambucus racemosa*

Table 2.48 Anatomical characters of *Sambucus* (Caprifoliaceae)

Character	*Sambucus nigra*	*Sambucus racemosa*
TS		
Vessel arrangement	semi-ring porous to diffuse porous	semi-ring porous to diffuse porous
Vessel grouping	clusters	clusters
Tyloses	present	present
Parenchyma	apotracheal	apotracheal
Ground tissue	thick-walled	thick-walled
Ring boundary	distinct ± 3 rows of vascular tracheids	distinct ± 10 rows of vascular tracheids
Ray/ring boundary	–	–
TLS		
Ray width	2–4	2–4
Ray height	10–30 (±20)	10–30 (±15)
Sheath cells	present	present
RLS		
Ray anatomy	heterogeneous types 3, 2 (1)	heterogeneous types 3, 2 (1)
Ray cell wall	thick-walled, beaded	thin-walled, not beaded
Libriform fibres	present	present
Fibre tracheids	absent	absent
Vascular tracheids	present	present
Spiral thickening	absent	absent
Perforation plates	simple	simple

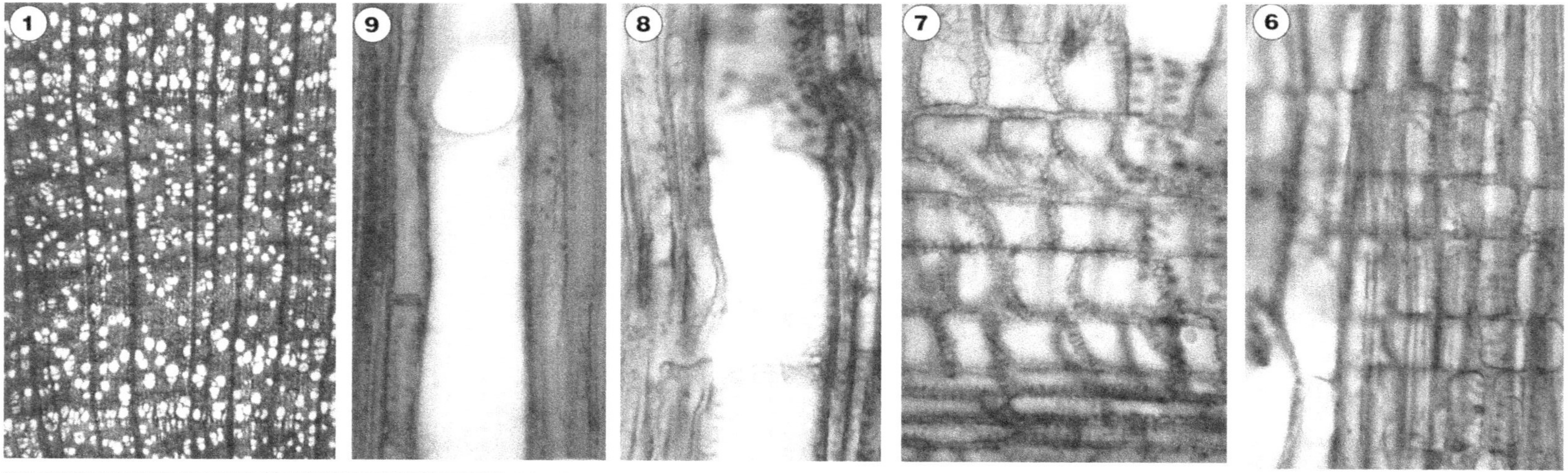

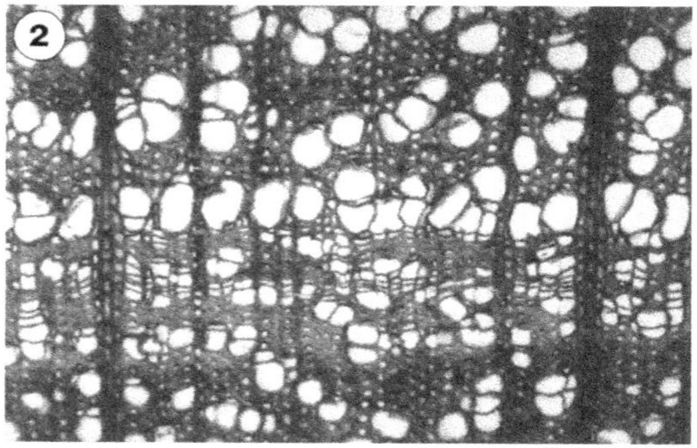

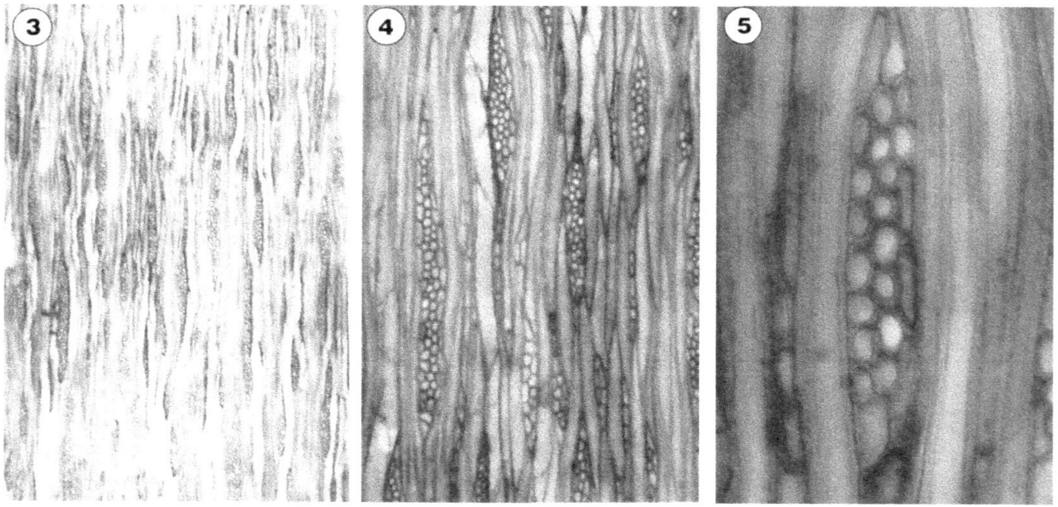

Plate 49 *Sambucus nigra* (1) TS (×28) (2) TS (×65) (3) TLS (×28) (4) TLS (×65) (5) TLS (×265) (6) RLS ray (×265) (7) RLS ray (×265) (8) RLS simple perforation plate (×265) (9) RLS simple perforation plate (×265)

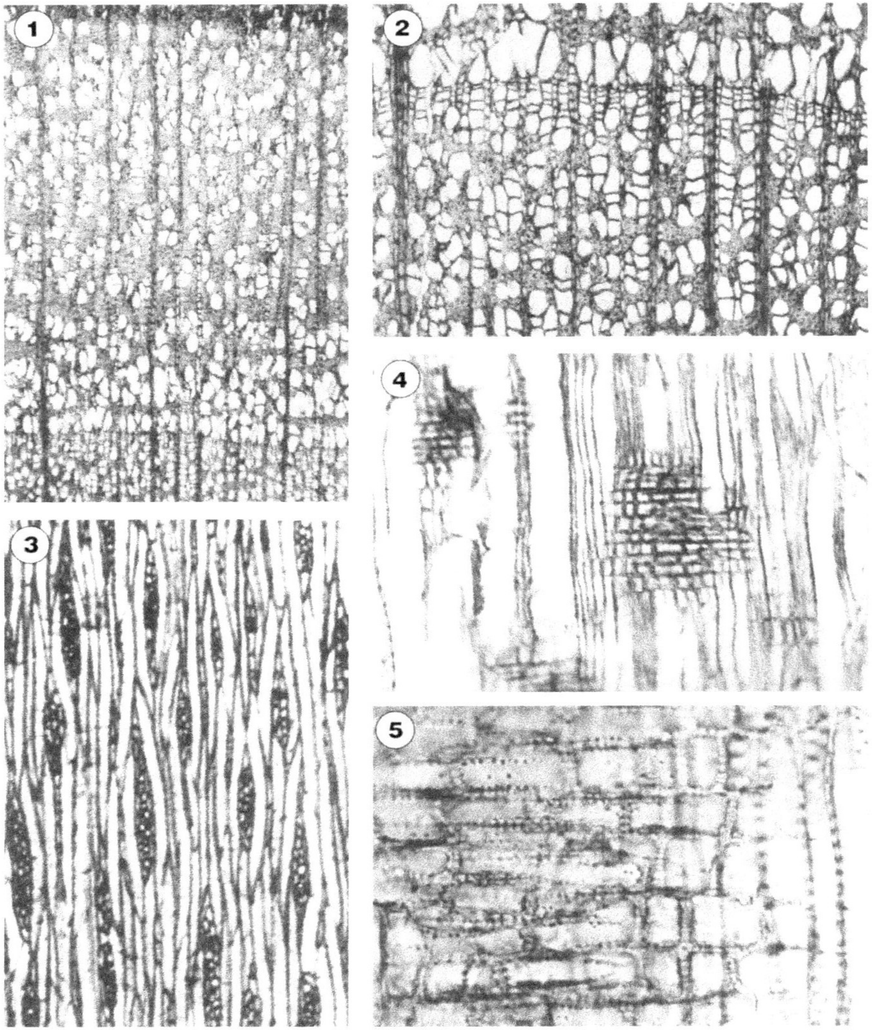

Plate 50 *Sambucus racemosa* (1) TS (×28) (2) TS (×65) (3) TLS (×65) (4) RLS (×65) (5) RLS ray (×265)

Hedera helix – Ivy (Araliaceae)

Plate 51

Hedera is a climbing plant of woodland margins with a western European distribution reaching into southern Scandinavia (see Map 122).

Hedera helix is diffuse porous with vessels distributed in more or less tangential clusters with paratracheal parenchyma. Apotracheal diffuse parenchyma is also present. Large rays may be distended at ring boundaries. Homogeneous and heterogeneous rays are present, up to 10 cells across and to 1mm high or more. In large rays, large irregular intercellular spaces form. Perforation plates are simple.

Hedera may be confused with the late wood of *Ulmus* spp. though the rays of *Ulmus* are much smaller.

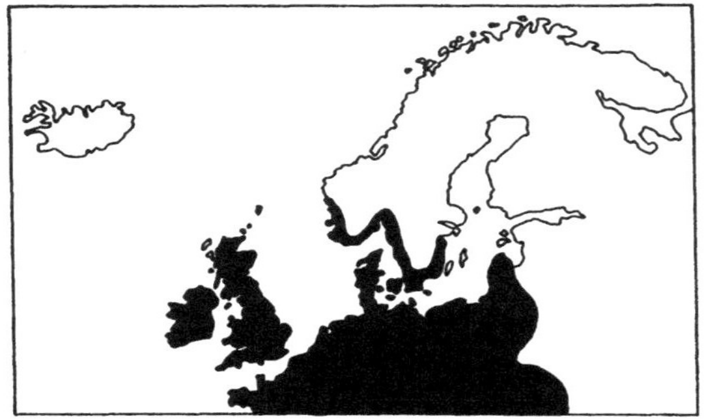

Map 122 Distribution of *Hedera helix*

Table 2.49 Anatomical characters of *Hedera helix* (Arailaceae)

Character	*Hedera helix*
TS	
Vessel arrangement	semi-ring porous to diffuse porous
Vessel grouping	tangential clusters
Tyloses	(present)
Parenchyma	apotracheal sparse, paratracheal
Ground tissue	in tangential groups between vessel clusters
Ring boundary	undulating
Ray/ring boundary	rays distended
TLS	
Ray width	1–10
Ray height	1mm+
Sheath cells	absent
RLS	
Ray anatomy	homogeneous, heterogeneous types 1,2,3
Libriform fibres	present (septate)
Fibre tracheids	absent
Vasicentric tracheids	present
Vascular tracheids	absent
Spiral thickening	absent
Perforation plates	simple

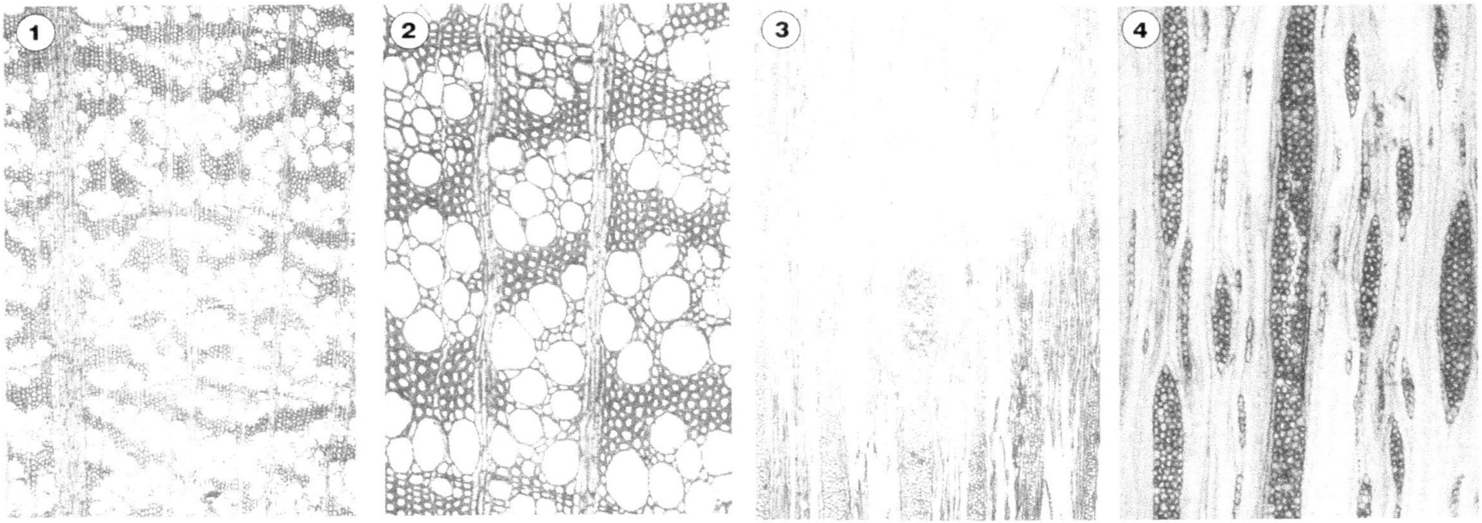

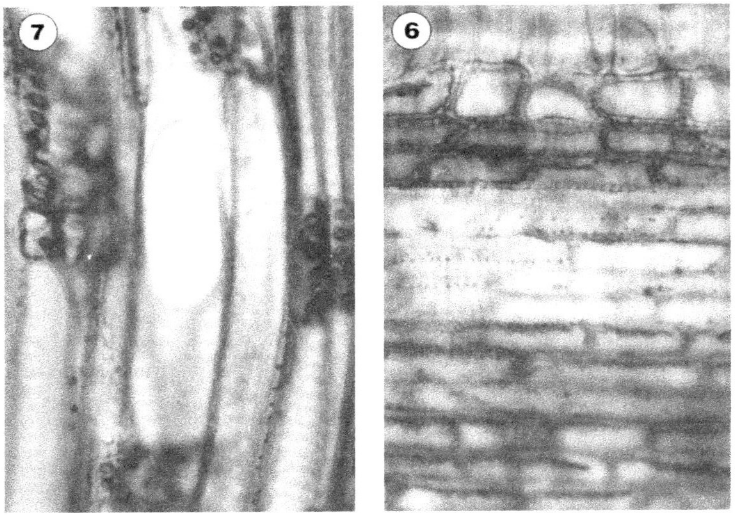

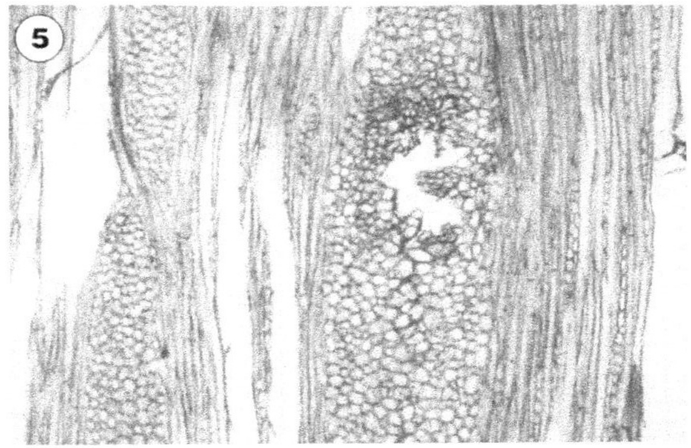

Plate 51 *Hedera helix* (1) TS (×28) (2) TS (×65) (3) TLS (×28) (4) TLS (×65) (5) TLS (×265) (6) RLS ray (×265) (7) RLS simple perforation plate (×265)

Group 8 – Diffuse porous woods with pores in dendritic arrangement

There are only two taxonomic groups with diffuse porous wood and a dendritic pore arrangement although there are three further ring-porous taxa with dendritic latewood that are dealt with below. All are small to medium shrubs or small trees.

General anatomy

A dendritic pore arrangement is where tangential or oblique bands of vessels are distributed across a year's growth in a net-like formation. These may be more or less regular or more random in appearance. The dendritic pattern may arise out of a continuous band of early wood vessels or may be initiated at the ring boundary. In the latter case the ring boundary may be indistinct. Only two taxonomic groups have this arrangement and so identification is usually not problematic although one group – the Leguminosae – is highly variable.

Table 2.50 Anatomical characters of woods with dendritic pore arrangement.

Taxon	Ray anatomy	Ray width	Ray height	Perforation plates	Spiral thickening
Leguminosae	Hm, Ht 3 (2)	3–7	–1mm	simple (scalariform)	present
Rhamnus catharticus	Ht 2–3	2–3	–10/15	simple	present – distinct

Three ring-porous woods have a late wood anatomy that is dendritic in pore arrangement, but in both *Quercus* and in *Castanea* the arrangement is rather weak and difficult to confuse with the taxa described here. *Berberis*, on the other hand, has similarities to *Rhamnus* in the distribution of vessels although the rays are very much larger. Possible confusion between *Berberis* and Leguminosae may occur – the height of rays in *Berberis* should separate the taxa. The rays in *Rhamnus* are also homogeneous whereas in both *Berberis* and Leguminosae heterogeneous rays are present.

Table 2.51 Anatomical characters of ring porous woods with dendritic late wood

Taxa/character state	Ray anatomy	Ray width	Ray height	Perforation plate	Spiral thickening
Quercus	Hm	1, 1mm	7–20, –5mm	simple	absent
Castanaea sativa	Hm (Ht 3)	1(2)	7–20	simple	absent
Berberis vulgaris	Hm	(1) 4–8	–1cm	simple	present

Taxa description

Group key

Rays under 20 cells high	*Rhamnus*
Rays over 20 cells high	*Leguminosae*

Rhamnus catharticus – Common Buckthorn (Rhamnaceae)

Plate 52

Rhamnus catharticus is a medium-sized thorny shrub of hedgerows and woodland margins. It has a distribution across most of northern Europe reaching into the southern part of Scandinavia (see Map 123).

Rhamnus catharticus has a diffuse porous distribution of vessels that decrease little in size over a year's growth. The vessels are solitary and in clusters grouped in a dendritic arrangement with a continuous band of vessels in the early wood. Because of this the ring boundary is very distinct. Paratracheal parenchyma is present. Rays are mostly bisereate with some trisereate rays and heterogeneous with one to two rows of upright cells that are larger than the body cells of the ray and can easily be seen in TLS. Rays are mostly up to 10 cells high. Perforation plates are simple and spiral thickening is present.

Rhamnus saxatalis, a closely related species, is similar in all respects apart from the ray height which, according to Schweingruber (1990), is up to 15 cells high. Although perhaps not common, rays of this height are present in *R. catharticus*. Differentiation in archaeological material is unlikely.

Confusion with Leguminosae is possible although the rays of the Leguminous woods in northern Europe are wider and much higher (see below).

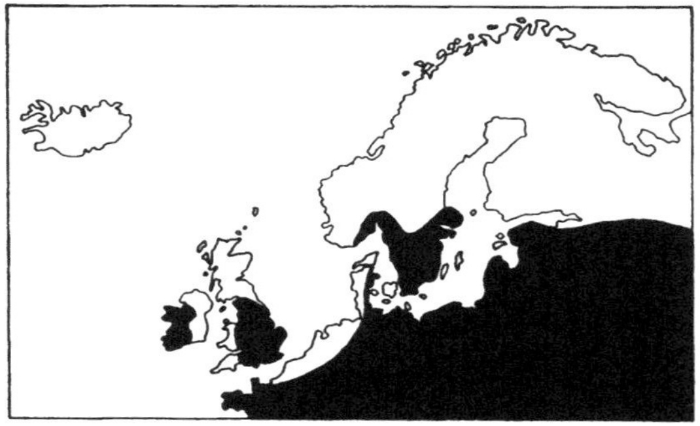

Map 123 Distribution of *Rhamnus catharticus*

Table 2.52 Anatomical characters of the *Rhamnus* (Rhamnaceae)

Character	*Rhamnus catharticus*	*Rhamnus saxatalis*	Character	*Rhamnus catharticus*	*Rhamnus saxatalis*
TS			RLS		
Vessel arrangement	diffuse porous	diffuse porous	Ray anatomy	heterogeneous types 2–3	heterogeneous types 2–3
Vessel grouping	solitary, clusters – dendritic	solitary, clusters – dendritic	Libriform fibres	present	present
Tyloses	present	present	Fibre tracheids	absent	absent
Parenchyma	paratracheal	paratracheal	Vascular tracheids	present	present
Ground tissue	–	–	Spiral thickening	distinct	distinct
Ring boundary	distinct	distinct	Crystals	present	present
Ray/ring boundary	–	–	Perforation plates	simple	simple
TLS					
Ray width	2–3	2–3			
Ray height	–10	–15			
Sheath cells	absent	absent			

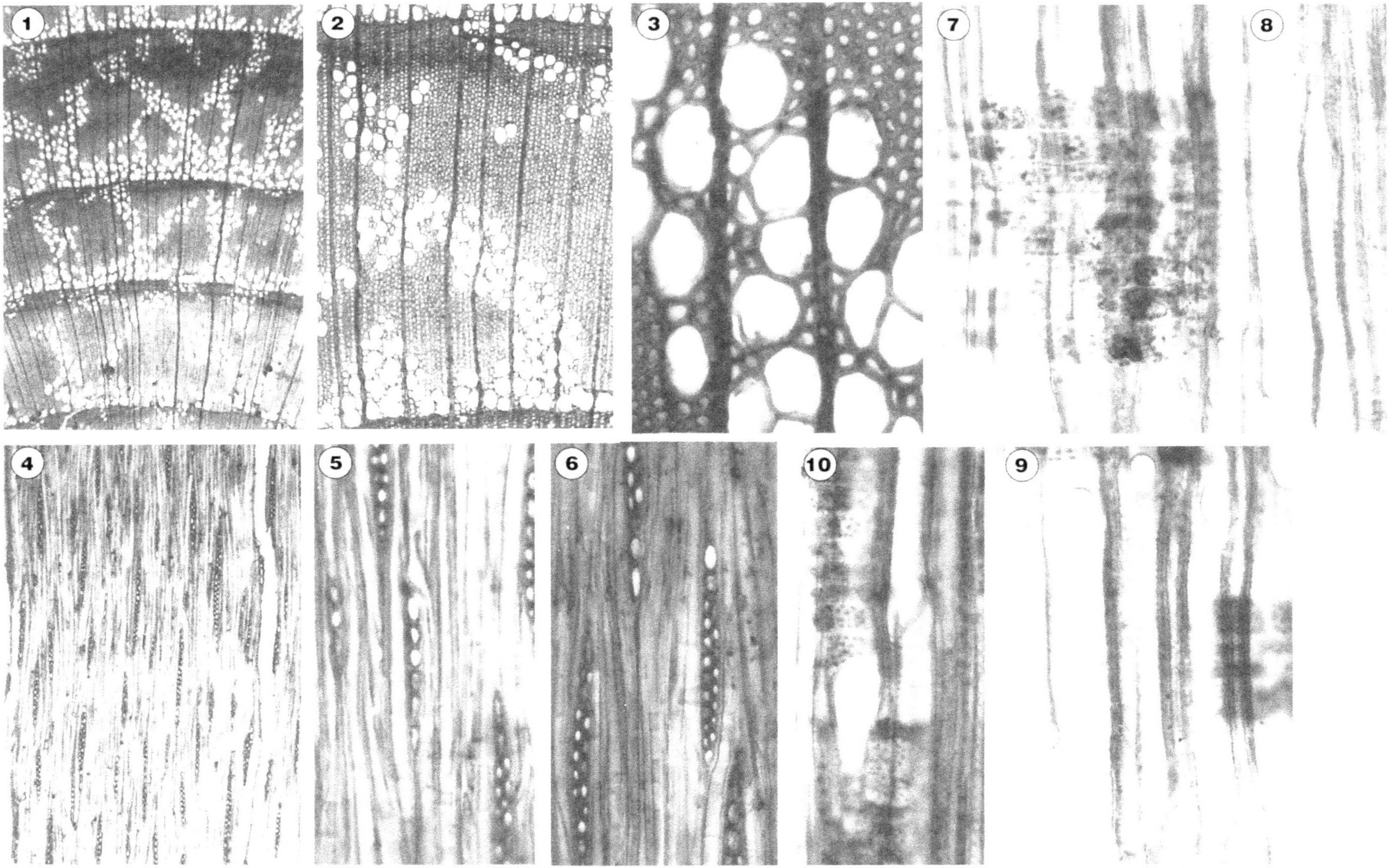

Plate 52 *Rhamnus catharticus* (1) TS (×28) (2) TS (×65) (3) TS (×265) (4) TLS (×65) (5) TLS (×265) (6) TLS (×265) (7) RLS ray (×265) (8) RLS simple perforation plate (×265) (9) RLS simple perforation plates (×265) (10) RLS simple perforation plate (×265).

Leguminosae (Fabaceae)

Plate 53

The woods of this family with a northern European distribution cannot be identified to the level of genus or species. They are all small to large shrubs with a mostly western northern European distribution (see Maps 124 to 131). In northern Europe the group includes:

Coloutea arborescens – bladder senna (Map 124)
Cytisus scoparius – broom (Map 125)
Genista pilosa – hairy greenweed (Map 126)
Genista anglica – petty whin (Map 127)
Genista tinctoria – Dyer's greenweed (Map 128)
Ulex europaeus – gorse (Map 129)
Ulex galii – Western gorse (Map 130)
Ulex minor – dwarf gorse (Map 131)

Unlike the Salicaceae, whose two genera and many species are difficult to differentiate as anatomically they are so similar, the genera and species of the Leguminosae are difficult to differentiate because the variation in the anatomy within one species may be as great as the variation within the family. The result of this is that the woods of two species may be very different or may be very similar.

Woods of the Leguminosae listed above are mostly diffuse porous to semi-ring porous, though in some Leguminous taxa a ring porous vessel form is evident. However, this is not common in the taxa examined and so the group is discussed here. The early wood vessels do not always form an entire ring and the dendritic arrangement of vessels is sometimes somewhat discontinuous. Vessels are solitary and in clusters with vascular tracheids and paratracheal parenchyma. Ground tissue is generally very thick-walled. Homogeneous and heterogeneous rays are present up to 7 cells wide and up to 1mm high. Parenchyma is fusiform and storied. Perforation plates are simple although occasional scalariform perforation plates are present in the late wood. Spiral thickening is present.

Confusion with *Rhamnus* is possible but the rays in *Rhamnus* are much smaller.

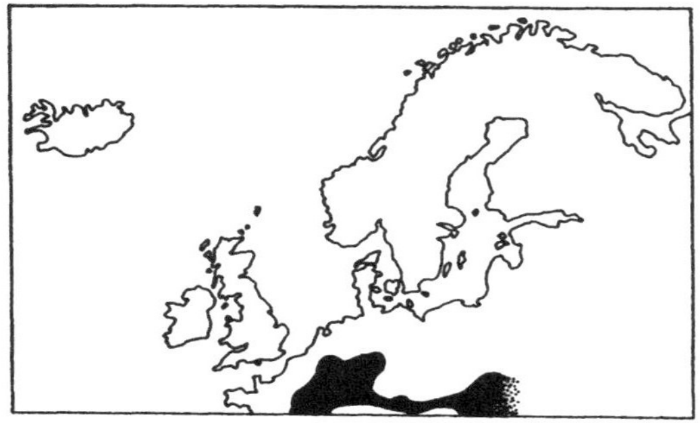

Map 124 Distribution of *Coloutea arborescens*

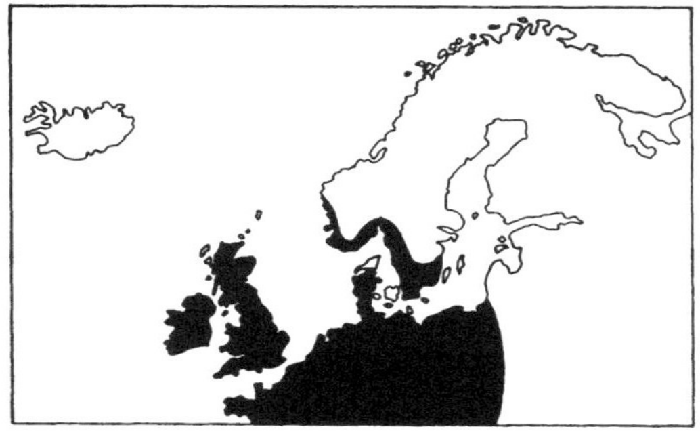

Map 125 Distribution of *Cytisus scoparius*

Map 126 Distribution of *Genista pilosa*

Table 2.53 Anatomical characters of the Leguminoseae (Fabaceae)

Character	Leguminosae
TS	
Vessel arrangement	ring porous/semi-ring porous
Vessel grouping	solitary, clusters – dendritic
Tyloses	present
Parenchyma	paratracheal
Ground tissue	thick-walled
Ring boundary	sometimes distinct
Ray/ring boundary	–
TLS	
Ray width	3–7
Ray height	–1mm
Sheath cells	present
RLS	
Ray anatomy	homogenous, heterogeneous types 3(2)
Libriform fibres	present
Fibre tracheids	absent
Vascular tracheids	(present)
Spiral thickening	present
Crystals	(present)
Perforation plates	simple

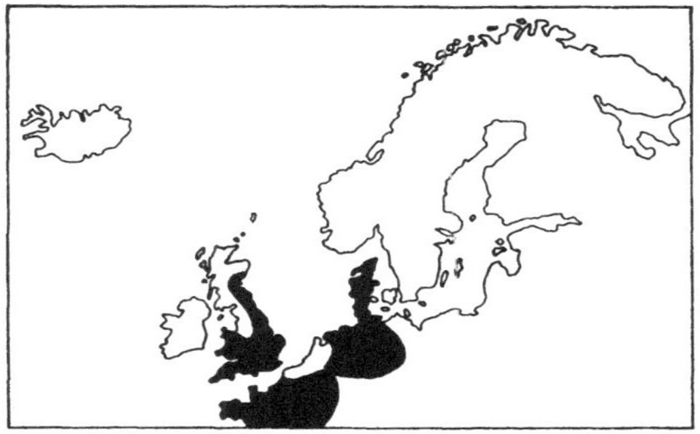

Map 127 Distribution of *Genista anglica*

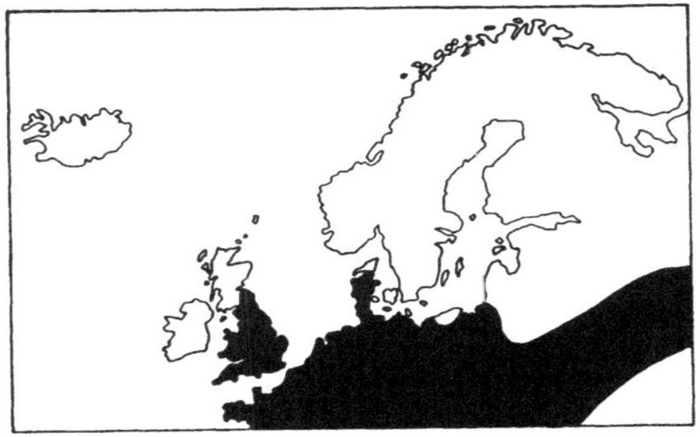

Map 128 Distribution of *Genista tinctoria*

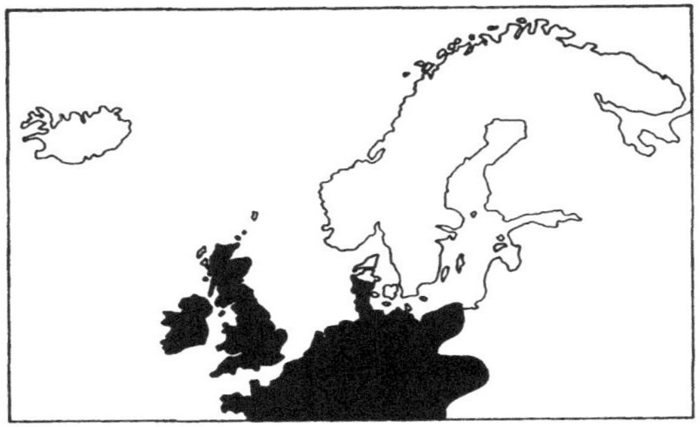

Map 129 Distribution of *Ulex europaeus*

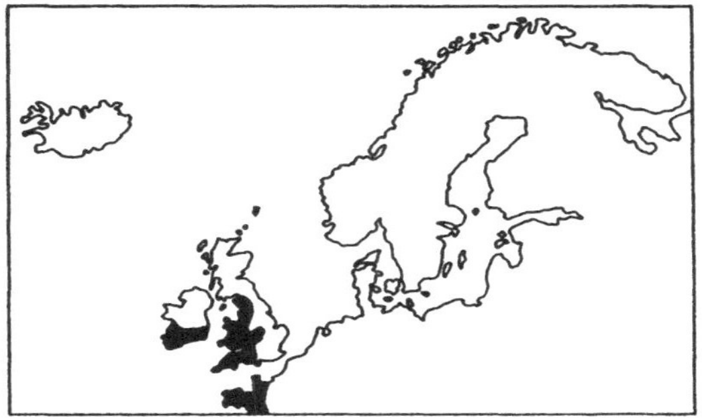

Map 130 Distribution of *Ulex galii*

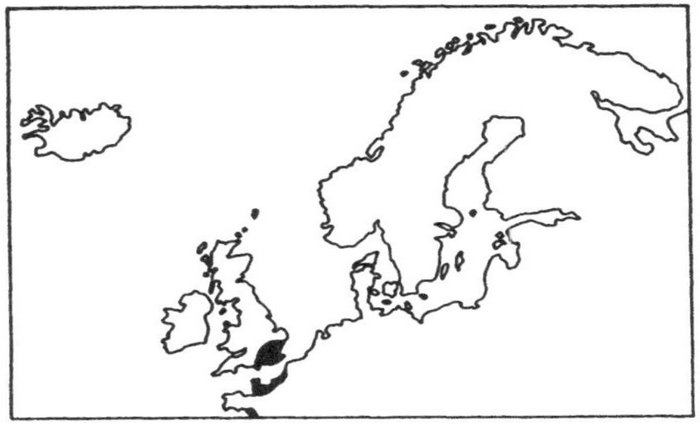

Map 131 Distribution of *Ulex minor*

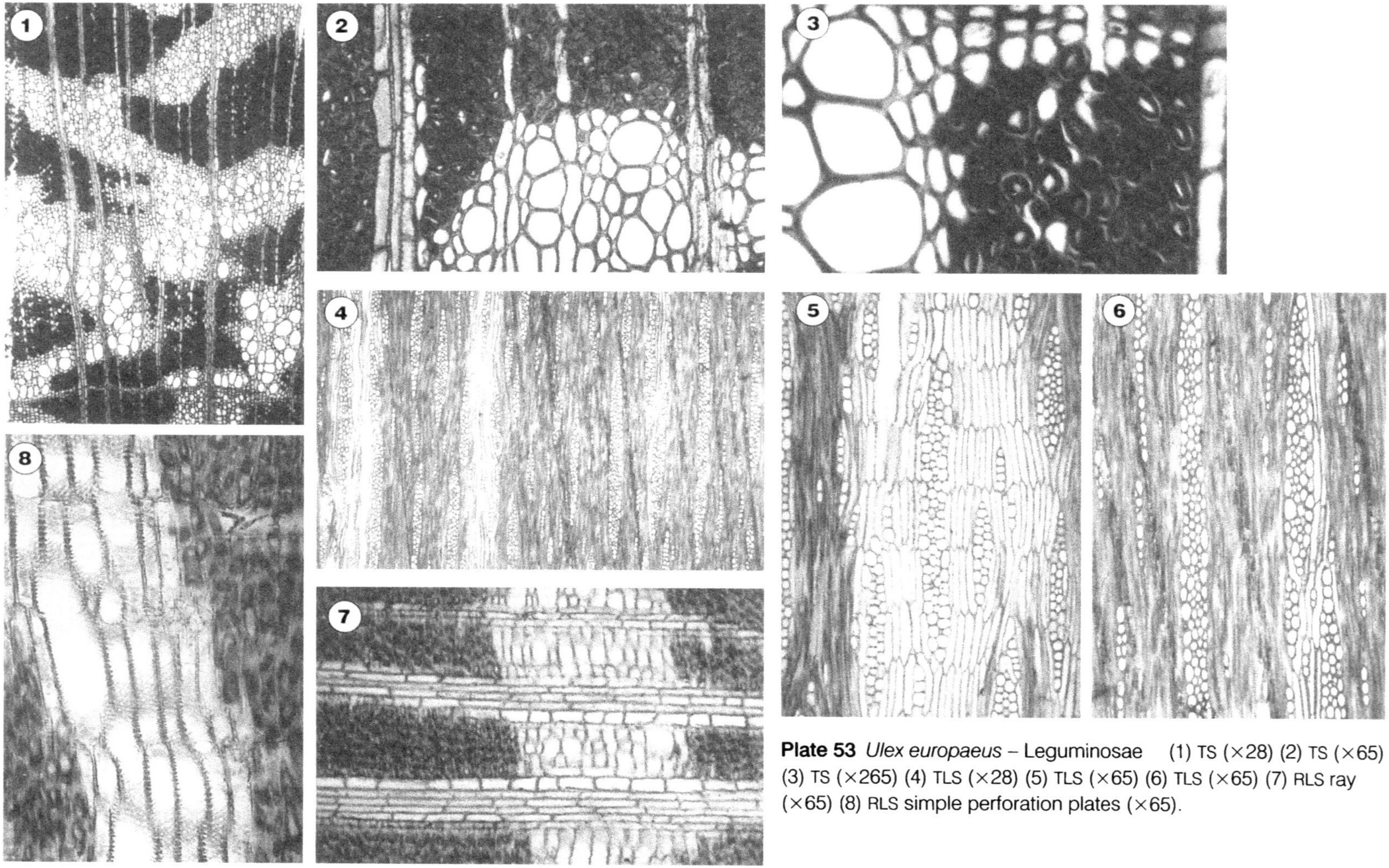

Plate 53 *Ulex europaeus* – Leguminosae (1) TS (×28) (2) TS (×65) (3) TS (×265) (4) TLS (×28) (5) TLS (×65) (6) TLS (×65) (7) RLS ray (×65) (8) RLS simple perforation plates (×65).

PART THREE

Character Tables

Introduction

The character tables that appear below are designed to aid identification of wood particularly where not all characters are easily visible. This is often the case in waterlogged wood which has become compressed or distorted under the weight of overburden and in charcoal. By picking a single character, such as ray size or anatomy, presence of spiral thickening or distribution of vessels in transverse section, all taxa that share the character can be seen. By checking each of these taxa in either these tables or in the descriptions in Part 2, other diagnostic characters can be either located or dismissed.

In all but one entry in these tables (*Taxus baccata* under prominent spiral thickening) the Gymnosperms are not included in these tables. There are relatively few Gymnosperms and the diagnostic characters used in their identification are mostly different to those used to identify the Dicotyledonous woods. The characters used to identify the Gymnosperms are dealt with fully on pages 30 to 43.

Table 3.1 RLS anatomy sorted by species

Taxon	Perforation plate (RLS)	Ray anatomy (RLS)	Vessel arrangement (TS)
Acer campestre, Acer opulus, monspessulanum	simple	homogeneous	DP solitary/files and clusters
Acer pseudoplatanis, T. platanoides	simple	homogeneous	DP solitary/files and clusters
Alnus glutinosa, A. incana	scalariform 10–20+	homogeneous	DP long radial files
Arbuteus unedo	simple (scalariform 1–4)	heterogeneous types 1 2 and 3	DP solitary/files and clusters
Berberis vulgaris	simple	homogeneous	ring porous
Betula pendula, B. pubescens	scalariform 10–25	homogeneous (heterogeneous type 3)	DP solitary/files and clusters
Buxus sempervirens	scalariform 5–12	heterogeneous type 3 (types 1 and 2)	DP solitary
Calluna vulgaris	simple (sclariform 1–5)	homogeneous	DP solitary
Carpinus betulus	simple	heterogeneous types 2 and 3 (homogeneous)	DP long radial files
Castanea sativa	simple	homogeneous (heterogeneous type 3)	ring porous
Clematis vitalba	simple	heterogeneous types 1, 2 and 3	ring porous
Cornus mas	scalariform 20+	heterogeneous type 1	DP solitary
Cornus sanguinea	scalariform 20+	heterogeneous type 1	DP solitary
Corylus avellana	scalariform 5–10	heterogeneous types 2 and 3 (homogeneous)	DP long radial files
Euonymus europaeus	simple	homogeneous (heterogeneous type 3)	DP solitary
Fagus sylvatica	simple (scalariform 8–15+)	homogeneous	DP solitary
Ficus carica	simple	heterogeneous types 2 and 3	DP solitary/files and clusters
Frangula alnus	simple	heterogeneous type 3	DP solitary/files and clusters
Fraxinus excelsior	simple	heterogeneous types 1, 2 and 3	ring porous
Hedera helix	simple	homogeneous, heterogeneous types 1, 2 and 3	DP clusters
Hippophae rhamnoides	simple	homogeneous (heterogeneous type 3)	ring porous
Ilex aquifolium	scalariform 10–30	heterogeneous type 1	DP long radial files
Juglans regia	simple	homogeneous (heterogeneous type 3)	DP solitary/files and clusters
Leguminosae	simple	homogeneous, heterogeneous type 3 (type 2)	DP dendritic
Ligustrum vulgare	simple	heterogeneous types 1, 2 and 3	DP solitary
Lonicera periclymenum	simple (scalariform 4–6)	heterogeneous type 1	DP solitary
Lonicera xylosteum, L. nigra	simple (scalariform 4–6)	heterogeneous type 1	DP solitary
Maloideae	simple	homogeneous (heterogeneous type 3)	DP solitary
Morus nigra	simple	heterogeneous types 2 and 3	ring porous
Populus spp.	simple	homogeneous (heterogeneous type 3)	DP solitary/files and clusters
Prunus avium	simple	homogeneous (heterogeneous type 3)	DP solitary/files and clusters

Taxon	Perforation plate (RLS)	Ray anatomy (RLS)	Vessel arrangement (TS)
Prunus padus	simple	homogeneous (heterogeneous type 3)	DP solitary/files and clusters
Prunus spinosa, P. domestica	simple	homogeneous (heterogeneous type 3)	DP solitary/files and clusters
Quercus petraea, Q. robur	simple	homogeneous	ring porous
Rhamnus catharticus	simple	heterogeneous types 2 and 3	DP dendritic
Rhamnus saxatalis	simple	heterogeneous types 2 and 3	DP dendritic
Rosa spp.	simple	heterogeneous type 1	ring porous
Rubus fruticosus agg. *Rubus idaeus*	simple	homogeneous, heterogeneous type 1	DP solitary/files and clusters
Sambucus nigra, S.racemosa	simple	heterogeneous types 2 and 3 (type 1)	DP clusters
Salix spp.	simple	heterogeneous type 3 (homogeneous)	DP solitary/files and clusters
Suaeda vera	simple	rays mostly absent	included phloem
Tilia cordata, T. platyphyllos	simple	homogeneous, heterogeneous type 3	
Ulmus glabra, Ulmus minor sl., *Ulmus laevis*	simple	homogeneous (heterogeneous type 3)	ring porous
Viburnum lantana	scalariform 25+	heterogeneous type 1	DP solitary
Viburnum opulus	scalariform 25+	heterogeneous type 1	DP solitary
Vitis vinifera	simple	homogeneous (heterogeneous types 2 and 3)	ring porous

Table 3.2 RLS anatomy sorted by perforation plate type

Taxon	Perforation plate (RLS)	Ray anatomy (RLS)	Vessel arrangement (TS)
Alnus glutinosa, A. incana	scalariform 10–20+	homogeneous	DP long radial files
Betula pendula, B. pubescens	scalariform 10–25	homogeneous (heterogeneous type 3)	DP solitary/files and clusters
Ilex aquifolium	scalariform 10–30	heterogeneous type 1	DP long radial files
Cornus mas	scalariform 20+	heterogeneous type 1	DP solitary
Cornus sanguinea	scalariform 20+	heterogeneous type 1	DP solitary
Viburnum lantana	scalariform 25+	heterogeneous type 1	DP solitary
Viburnum opulus	scalariform 25+	heterogeneous type 1	DP solitary
Corylus avellana	scalariform 5–10	heterogeneous types 2 and 3 (homogeneous)	DP long radial files
Buxus sempervirens	scalariform 5–12	heterogeneous type 3 (types 1 and 2)	DP solitary
Arbuteus unedo	simple (scalariform 1–4)	heterogeneous types 1, 2 and 3	DP solitary/files and clusters
Lonicera xylosteum, L. nigra	simple (scalariform 4–6)	heterogeneous type 1	DP solitary
Lonicera periclymenum	simple (scalariform 4–6)	heterogeneous type 1	DP solitary
Fagus sylvatica	simple (scalariform 8–15+)	homogeneous	DP solitary
Calluna vulgaris	simple (sclariform 1–5)	homogeneous	DP solitary

Taxon	Perforation plate (RLS)	Ray anatomy (RLS)	Vessel arrangement (TS)
Acer campestre, Acer opulus, monspessulanum	simple	homogeneous	DP solitary/files and clusters
Acer pseudoplatanis, A. platanoides	simple	homogeneous	DP solitary/files and clusters
Maloideae	simple	homogeneous (heterogeneous type 3)	DP solitary
Berberis vulgaris	simple	homogeneous	ring porous
Carpinus betulus	simple	heterogeneous types 2 and 3 (homogeneous)	DP long radial files
Castanea sativa	simple	homogeneous (heterogeneous type 3)	ring porous
Clematis vitalba	simple	heterogeneous types 1, 2 and 3	ring porous
Euonymus europaeus	simple	homogeneous (heterogeneous type 3)	DP solitary
Ficus carica	simple	heterogeneous types 2 and 3	DP solitary/files and clusters
Frangula alnus	simple	heterogeneous type 3	DP solitary/files and clusters
Fraxinus excelsior	simple	heterogeneous types 1, 2 and 3	ring porous
Hedera helix	simple	homogeneous, heterogeneous types 1, 2 and 3	DP clusters
Hippophae rhamnoides	simple	homogeneous (heterogeneous type 3)	ring porous
Juglans regia	simple	homogeneous (heterogeneous type 3)	DP solitary/files and clusters
Leguminosae	simple	homogeneous, heterogeneous type 3 (type 2)	DP dendritic
Ligustrum vulgare	simple	heterogeneous types 1, 2 and 3	DP solitary
Morus nigra	simple	heterogeneous types 2 and 3	ring porous
Populus spp.	simple	homogeneous (heterogeneous type 3)	DP solitary/files and clusters
Prunus avium	simple	homogeneous (heterogeneous type 3)	DP solitary/files and clusters
Prunus padus	simple	homogeneous (heterogeneous type 3)	DP solitary/files and clusters
Prunus spinosa, P. domestica	simple	homogeneous (heterogeneous type 3)	DP solitary/files and clusters
Quercus petraea, Q. robur	simple	homogeneous	ring porous
Rhamnus catharticus	simple	heterogeneous types 2 and 3	DP dendritic
Rhamnus saxatalis	simple	heterogeneous types 2 and 3	DP dendritic
Rosa spp.	simple	heterogeneous type 1	ring porous
Rubus fruticosus agg. *Rubus idaeus*	simple	homogeneous, heterogeneous type 1	DP solitary/files and clusters
Sambucus nigra, S. racemosa	simple	heterogeneous types 2 and 3 (type 1)	DP clusters
Salix spp.	simple	heterogeneous type 3 (homogeneous)	DP solitary/files and clusters
Suaeda vera	simple	rays mostly absent	included phloem
Tilia cordata, T. platyphyllos	simple	homogeneous, heterogeneous type 3	DP solitary/files and clusters
Ulmus glabra, Ulmus minor sl., *Ulmus laevis*	simple	homogeneous (heterogeneous type 3)	ring porous
Vitis vinifera	simple	homogeneous (heterogeneous types 2 and 3)	ring porous

Table 3.3 RLS anatomy sorted by ray anatomy

Taxon	Perforation plate (RLS)	Ray anatomy (RLS)	Vessel arrangement (TS)
Alnus glutinosa, A. incana	scalariform 10–20+	homogeneous	DP long radial files
Fagus sylvatica	simple (scalariform 8–15+)	homogeneous	DP solitary
Calluna vulgaris	simple (sclariform 1–5)	homogeneous	DP solitary
Acer campestre, Acer opulus, monspessulanum	simple	homogeneous	DP solitary/files and clusters
Acer pseudoplatanis, A. platanoides	simple	homogeneous	DP solitary/files and clusters
Berberis vulgaris	simple	homogeneous	ring porous
Quercus petraea, Q. robur	simple	homogeneous	ring porous
Betula pendula, B. pubescens	scalariform 10–25	homogeneous (heterogeneous type 3)	DP solitary/files and clusters
Maloideae	simple	homogeneous (heterogeneous type 3)	DP solitary
Castanea sativa	simple	homogeneous (heterogeneous type 3)	ring porous
Euonymus europaeus	simple	homogeneous (heterogeneous type 3)	DP solitary
Hippophae rhamnoides	simple	homogeneous (heterogeneous type 3)	ring porous
Juglans regia	simple	homogeneous (heterogeneous type 3)	DP solitary/files and clusters
Populus spp.	simple	homogeneous (heterogeneous type 3)	DP solitary/files and clusters
Prunus avium	simple	homogeneous (heterogeneous type 3)	DP solitary/files and clusters
Prunus padus	simple	homogeneous (heterogeneous type 3)	DP solitary/files and clusters
Prunus spinosa, P. domestica	simple	homogeneous (heterogeneous type 3)	DP solitary/files and clusters
Ulmus glabra, Ulmus minor sl., *Ulmus laevis*	simple	homogeneous (heterogeneous type 3)	ring porous
Vitis vinifera	simple	homogeneous (heterogeneous types 2 and 3)	ring porous
Rubus fruticosus agg. *Rubus idaeus*	simple	homogeneous, heterogeneous type 1	DP solitary/files and clusters
Tilia cordata, platyphyllos	simple	homogeneous, heterogeneous type 3	DP solitary/files and clusters
Leguminosae	simple	homogeneous, heterogeneous type 3 (type 2)	DP dendritic
Hedera helix	simple	homogeneous, heterogeneous types 1, 2 and 3	DP clusters
Ilex aquifolium	scalariform 10–30	heterogeneous type 1	DP long radial files
Cornus mas	scalariform 20+	heterogeneous type 1	DP solitary
Cornus sanguinea	scalariform 20+	heterogeneous type 1	DP solitary
Viburnum lantana	scalariform 25+	heterogeneous type 1	DP solitary
Viburnum opulus	scalariform 25+	heterogeneous type 1	DP solitary
Lonicera xylosteum, L. nigra	simple (scalariform 4–6)	heterogeneous type 1	DP solitary
Lonicera periclymenum	simple (scalariform 4–6)	heterogeneous type 1	DP solitary
Rosa spp.	simple	heterogeneous type 1	ring porous

Taxon	Perforation plate (RLS)	Ray anatomy (RLS)	Vessel arrangement (TS)
Frangula alnus	simple	heterogeneous type 3	DP solitary/files and clusters
Salix spp.	simple	heterogeneous type 3 (homogeneous)	DP solitary/files and clusters
Buxus sempervirens	scalariform 5–12	heterogeneous type 3 (types 1 and 2)	DP solitary
Ficus carica	simple	heterogeneous types 2 and 3	DP solitary/files and clusters
Morus nigra	simple	heterogeneous types 2 and 3	ring porous
Arbuteus unedo	simple (scalariform 1–4)	heterogeneous types 1, 2 and 3	DP solitary/files and clusters
Clematis vitalba	simple	heterogeneous types 1, 2 and 3	ring porous
Fraxinus excelsior	simple	heterogeneous types 1, 2 and 3	ring porous
Ligustrum vulgare	simple	heterogeneous types 1, 2 and 3	DP solitary
Rhamnus catharticus	simple	heterogeneous types 2 and 3	DP dendritic
Rhamnus saxatalis	simple	heterogeneous types 2 and 3	DP dendritic
Corylus avellana	scalariform 5–10	heterogeneous types 2 and 3 (homogeneous)	DP long radial files
Carpinus betulus	simple	heterogeneous types 2 and 3 (homogeneous)	DP long radial files
Sambucus nigra, S. racemosa	simple	heterogeneous types 2 and 3 (type 1)	DP clusters
Suaeda vera	simple	rays mostly absent	included phloem

Table 3.4 RLS anatomy sorted by TS anatomy

Taxon	Perforation plate (RLS)	Ray anatomy (RLS)	Vessel arrangement (TS)
Berberis vulgaris	simple	homogeneous	ring porous
Castanea sativa	simple	homogeneous (heterogeneous type 3)	ring porous
Clematis vitalba	simple	heterogeneous types 1, 2 and 3	ring porous
Fraxinus excelsior	simple	heterogeneous types 1, 2 and 3	ring porous
Hippophae rhamnoides	simple	homogeneous (heterogeneous type 3)	ring porous
Morus nigra	simple	heterogeneous types 2 and 3	ring porous
Quercus petraea, Q. robur	simple	homogeneous	ring porous
Rosa spp.	simple	heterogeneous type 1	ring porous
Ulmus glabra, Ulmus minor sl., *Ulmus laevis*	simple	homogeneous (heterogeneous type 3)	ring porous
Vitis vinifera	simple	homogeneous (heterogeneous types 2 and 3)	ring porous
Suaeda vera	simple	rays mostly absent	included phloem
Tilia cordata, T. platyphyllos	simple	homogeneous, heterogeneous type 3	DP solitary/files and clusters
Betula pendula, B. pubescens	scalariform 10–25	homogeneous (heterogeneous type 3)	DP solitary/files and clusters
Arbuteus unedo	simple (scalariform 1–4)	heterogeneous types 1, 2 and 3	DP solitary/files and clusters

Taxon	Perforation plate (RLS)	Ray anatomy (RLS)	Vessel arrangement (TS)
Acer campestre, Acer opulus, monspessulanum	simple	homogeneous	DP solitary/files and clusters
Acer pseudoplatanis, A. platanoides	simple	homogeneous	DP solitary/files and clusters
Ficus carica	simple	heterogeneous types 2 and 3	DP solitary/files and clusters
Frangula alnus	simple	heterogeneous type 3	DP solitary/files and clusters
Juglans regia	simple	homogeneous (heterogeneous type 3)	DP solitary/files and clusters
Populus spp.	simple	homogeneous (heterogeneous type 3)	DP solitary/files and clusters
Prunus avium	simple	homogeneous (heterogeneous type 3)	DP solitary/files and clusters
Prunus padus	simple	homogeneous (heterogeneous type 3)	DP solitary/files and clusters
Prunus spinosa, P. domestica	simple	homogeneous (heterogeneous type 3)	DP solitary/files and clusters
Rubus fruticosus agg. *Rubus idaeus*	simple	homogeneous, heterogeneous type 1	DP solitary/files and clusters
Salix spp.	simple	heterogeneous type 3 (homogeneous)	DP solitary/files and clusters
Cornus mas	scalariform 20+	heterogeneous type 1	DP solitary
Cornus sanguinea	scalariform 20+	heterogeneous type 1	DP solitary
Viburnum lantana	scalariform 25+	heterogeneous type 1	DP solitary
Viburnum opulus	scalariform 25+	heterogeneous type 1	DP solitary
Buxus sempervirens	scalariform 5–12	heterogeneous type 3 (types 1 and 2)	DP solitary
Lonicera xylosteum, L. nigra	simple (scalariform 4–6)	heterogeneous type 1	DP solitary
Lonicera periclymenum	simple (scalariform 4–6)	heterogeneous type 1	DP solitary
Fagus sylvatica	simple (scalariform 8–15+)	homogeneous	DP solitary
Calluna vulgaris	simple (sclariform 1–5)	homogeneous	DP solitary
Maloideae	simple	homogeneous (heterogeneous type 3)	DP solitary
Euonymus europaeus	simple	homogeneous (heterogeneous type 3)	DP solitary
Ligustrum vulgare	simple	heterogeneous types 1, 2 and 3	DP solitary
Alnus glutinosa, A. incana	scalariform 10–20+	homogeneous	DP long radial files
Ilex aquifolium	scalariform 10–30	heterogeneous type 1	DP long radial files
Corylus avellana	scalariform 5–10	heterogeneous types 2 and 3 (homogeneous)	DP long radial files
Carpinus betulus	simple	heterogeneous types 2 and 3 (homogeneous)	DP long radial files
Leguminosae	simple	homogeneous, heterogeneous type 3 (type 2)	DP dendritic
Rhamnus catharticus	simple	heterogeneous types 2 and 3	DP dendritic
Rhamnus saxatalis	simple	heterogeneous types 2 and 3	DP dendritic
Hedera helix	simple	homogeneous, heterogeneous types 1, 2 and 3	DP clusters
Sambucus nigra, S.racemosa	simple	heterogeneous types 2 and 3 (type 1)	DP clusters

Table 3.5 TLS anatomy sorted by TS anatomy

Taxon	Ray anatomy (TLS)		Vessel arrangement (TS)
Berberis vulgaris	width: (1)4–8	height: –1 cm	ring porous
Castanea sativa	width: 1(2)	height: 7–20	ring porous
Clematis vitalba	width: 5–15+	height: 1 cm+	ring porous
Fraxinus excelsior	width: (1–)2–3	height: 10–20	ring porous
Hippophae rhamnoides	width: 1–2(–3)	height: 5–15	ring porous
Morus nigra	width: 4–7	height: –60	ring porous
Quercus petraea, Q. robur	width: 1, 1 mm	height: 7–20, –5 mm	ring porous
Rosa spp.	width: 1, 6–15	height: 3–12, –10mm	ring porous
Ulmus glabra, Ulmus minor sl., *Ulmus laevis*	width: 1, 4–5+	height: 10–50	ring porous
Vitis vinifera	width: 5–20	height: –2mm	ring porous
Acer campestre, Acer opulus, Acer monspessulanum	width: (1), 2–3(–4)	height: (5+) (20–30)	DP solitary/files and clusters
Acer pseudoplatanis, A. platanoides	width: (1–2), 3–6	height: (5+) (20–70)	DP solitary/files and clusters
Arbuteus unedo	width: 1, 2–3(–4)	height: 2–3, 5–15(–20)	DP solitary/files and clusters
Betula pendula, B. pubescens	width: 2–4(–5)	height: 10–15	DP solitary/files and clusters
Ficus carica	width: 3–4	height: –30	DP solitary/files and clusters
Frangula alnus	width: 1–3	height: 10–20	DP solitary/files and clusters
Juglans regia	width: (1–)2–4(–5)	height: 15–30	DP solitary/files and clusters
Populus spp.	width: 1	height: 10–15	DP solitary/files and clusters
Prunus avium	width: 2–4	height: –30	DP solitary/files and clusters
Prunus padus	width: 2–7	height: –30	DP solitary/files and clusters
Prunus spinosa, P. domestica	width: 3–7	height: 50+	DP solitary/files and clusters
Rubus fruticosus agg. *Rubus idaeus*	width: 1, 4–12	height: –12, –10 mm	DP solitary/files and clusters
Salix spp.	width: 1	height: 10–15	DP solitary/files and clusters
Tilia cordata, T. platyphyllos	width: 1, 2–4+	height: 5–10 (–20)	DP solitary/files and clusters
Buxus sempervirens	width: (1–)2	height: 6–12	DP solitary
Calluna vulgaris	width: 1	height: 1–5	DP solitary
Cornus mas	width: 1, 3–5(–7)	height: 5–50	DP solitary
Cornus sanguinea	width: 1, 3–5(–7)	height: 5–50	DP solitary
Euonymus europaeus	width: 1	height: 5–15	DP solitary
Fagus sylvatica	width: 1, –20	height: –5 mm	DP solitary
Ligustrum vulgare	width: 1–2(–3)	height: 10–20	DP solitary
Lonicera periclymenum	width: 1–3(–4)	height: 8–15	DP solitary
Lonicera xylosteum, L. nigra	width: 1–3(–4)	height: 8–15	DP solitary
Maloideae	width: 2–3	height: 5–15	DP solitary

Taxon	Ray anatomy (TLS)		Vessel arrangement (TS)
Viburnum lantana	width: 1–3	height: 5–15	DP solitary
Viburnum opulus	width: 1–3	height: 5–15	DP solitary
Alnus glutinosa, A. incana	width: 1, 1–3 (AR)	height: 3–25	DP long radial files
Carpinus betulus	width: 1–2 1–4 (AR)	height: 10–20+	DP long radial files
Corylus avellana	width: 1(–2) 1–3 (AR)	height: 10–25 (–40)	DP long radial files
Ilex aquifolium	width: 1, 4–8	height: 2–8, -3mm	DP long radial files
Leguminosae	width: 3–7	height: -1 mm	DP dendritic
Rhamnus catharticus	width: 2–3	height: –10	DP dendritic
Rhamnus saxatalis	width: 2–3	height: –15	DP dendritic
Hedera helix	width: 1–10	height: 1 mm+	DP clusters
Sambucus nigra, S. racemosa	width: 2–4	height: 10–30	DP clusters
Suaeda vera	rays mostly absent	included phloem	

Table 3.6 TLS anatomy sorted by species

Taxon	Ray anatomy (TLS)		Vessel arrangement (TS)
Acer campestre, Acer opulus, Acer monspessulanum	width: (1), 2–3(–4)	height: (5+) (20–30)	DP solitary/files and clusters
Acer pseudoplatanis, A. platanoides	width: (1–2), 3–6	height: (5+) (20–70)	DP solitary/files and clusters
Alnus glutinosa, A. incana	width: 1, 1–3 (AR)	height: 3–25	DP long radial files
Arbuteus unedo	width: 1, 2–3(–4)	height: 2–3, 5–15(–20)	DP solitary/files and clusters
Berberis vulgaris	width: (1)4–8	height: –1 cm	ring porous
Betula pendula, B. pubescens	width: 2–4(–5)	height: 10–15	DP solitary/files and clusters
Buxus sempervirens	width: (1–)2	height: 6–12	DP solitary
Calluna vulgaris	width: 1	height: 1–5	DP solitary
Carpinus betulus	width: 1–2 1–4 (AR)	height: 10–20+	DP long radial files
Castanea sativa	width: 1(2)	height: 7–20	ring porous
Clematis vitalba	width: 5–15+	height: 1cm+	ring porous
Cornus mas	width: 1, 3–5(–7)	height: 5–50	DP solitary
Cornus sanguinea	width: 1, 3–5(–7)	height: 5–50	DP solitary
Corylus avellana	width: 1(–2) 1–3 (AR)	height: 10–25 (–40)	DP long radial files
Euonymus europaeus	width: 1	height: 5–15	DP solitary
Fagus sylvatica	width: 1, –20	height: –5 mm	DP solitary
Ficus carica	width: 3–4	height: –30	DP solitary/files and clusters
Frangula alnus	width: 1–3	height: 10–20	DP solitary/files and clusters

Taxon	Ray anatomy (TLS)		Vessel arrangement (TS)
Fraxinus excelsior	width: (1–)2–3	height: 10–20	ring porous
Hedera helix	width: 1–10	height: 1 mm+	DP clusters
Hippophae rhamnoides	width: 1–2(–3)	height: 5–15	ring porous
Ilex aquifolium	width: 1, 4–8	height: 2–8, –3 mm	DP long radial files
Juglans regia	width: (1–)2–4(–5)	height: 15–30	DP solitary/files and clusters
Leguminosae	width: 3–7	height: –1 mm	DP dendritic
Ligustrum vulgare	width: 1–2(–3)	height: 10–20	DP solitary
Lonicera periclymenum	width: 1–3(–4)	height: 8–15	DP solitary
Lonicera xylosteum,L. nigra	width: 1–3(–4)	height: 8–15	DP solitary
Maloideae	width: 2–3	height: 5–15	DP solitary
Morus nigra	width: 4–7	height: –60	ring porous
Populus spp.	width: 1	height: 10–15	DP solitary/files and clusters
Prunus avium	width: 2–4	height: –30	DP solitary/files and clusters
Prunus padus	width: 2–7	height: –30	DP solitary/files and clusters
Prunus spinosa, P. domestica	width: 3–7	height: 50+	DP solitary / files and clusters
Quercus petraea, Q. robur	width: 1, 1 mm	height: 7–20, –5 mm	ring porous
Rhamnus catharticus	width: 2–3	height: –10	DP dendritic
Rhamnus saxatalis	width: 2–3	height: –15	DP dendritic
Rosa spp.	width: 1, 6–15	height: 3–12, –10 mm	ring porous
Rubus fruticosus agg. *Rubus idaeus*	width: 1, 4–12	height: –12, –10 mm	DP solitary/files and clusters
Sambucus nigra, S. racemosa	width: 2–4	height: 10–30	DP clusters
Salix spp.	width: 1	height: 10–15	DP solitary/files and clusters
Suaeda vera	rays mostly absent	included phloem	
Tilia cordata, platyphyllos	width: 1, 2–4+	height: 5–10 (–20)	DP solitary/files and clusters
Ulmus glabra, Ulmus minor sl., *Ulmus laevis*	width: 1, 4–5+	height: 10–50	ring porous
Viburnum lantana	width: 1–3	height: 5–15	DP solitary
Viburnum opulus	width: 1–3	height: 5–15	DP solitary
Vitis vinifera	width: 5–20	height: –2 mm	ring porous

Table 3.7 Unisereate rays present (TLS) (* = unisereate rays present only)

Taxon	Unisereate rays present (TLS)	Vessel arrangement (TS)
Acer	(present)	DP solitary, files and clusters
Alnus	(present)	DP long radial files
Arbuteus	present	DP solitary, files and clusters
Berberis	(present)	ring porous
Buxus	(present)	DP solitary
Calluna	present *	DP solitary
Carpinus	present	DP long radial files
Castanea	present	ring porous
Cornus	present	DP solitary
Corylus	present	DP long radial files
Euonymus	present *	DP solitary
Fagus	present	DP solitary
Frangula	(present)	DP solitary, files and clusters
Fraxinus	present	ring porous
Hedera	present	DP clusters
Hippophae	present	ring porous
Ilex	present	DP long radial files
Juglans	(present)	DP solitary, files and clusters
Ligustrum	present	DP solitary
Lonicera	present	DP solitary
Salicaceae	present *	DP solitary, files and clusters
Quercus	present	ring porous
Rosa	present	ring porous
Rubus	present	DP solitary, files and clusters
Tilia	present	DP solitary, files and clusters
Ulmus	present	ring porous
Viburnum	present	DP solitary

Table 3.8 Unisereate rays absent (TLS)

Taxon	Unisereate rays absent (TLS)	Vessel arrangement (TS)
Betula	absent	DP solitary, files and clusters
Clematis	absent	ring porous
Ficus	absent	DP solitary, files and clusters
Leguminosae	absent	DP dendritic
Maloideae	absent	DP solitary
Morus	absent	ring porous
Prunus	absent	DP solitary, files and clusters
Rhamnus	absent	DP dendritic
Sambucus	absent	DP clusters
Vitis	absent	ring porous

Table 3.9 Unisereate rays composed of upright cells only (TLS) (* = unisereate rays present only)

Taxon	Unisereate rays composed of upright cells only (TLS)	Vessel arrangement (TS)
Calluna	present * up to 5 cells high	DP solitary
Viburnum	present up to 10 cells high	DP solitary
Cornus	present up to 8 cells high	DP solitary
Rubus	present up to 12 cells high	DP solitary, files and clusters
Tilia	present up to 7 cells high	DP solitary, files and clusters
Arbutus	present up to 5 cells high	DP solitary, files and clusters
Ilex	present up to 8 cells high	DP long radial files

Table 3.10 Rays of two distinct sizes present (TLS)

Taxon	Rays of two distinct sizes present (TLS)		Vessel arrangement (TS)
Berberis	width: (1), 4–8	height: –1 cm	ring porous
Fagus	width: 1, –20	height: –5 mm	DP solitary
Ilex	width: 1, 4–8	height: –3 mm	DP long radial files
Quercus	width: 1, 1 mm	height: –5 mm	ring porous
Rosa	width: 1, 6–15	height: –10 mm	ring porous
Rubus	width: 1, 4–12	height: –10 mm	DP solitary, files and clusters
Ulmus	width: 1, 4–5+	height: 10–50	ring porous

Table 3.11 Large multisereate rays present (TLS)

Taxon	Large multisereate rays present (TLS)	Vessel arrangement (TS)
Acer pseudoplatanus *Acer platanoides*	3–6 20–70	DP solitary, files and clusters
Berberis	4–8 –1 cm	ring porous
Clematis	5–15 –1 cm	ring porous
Fagus	–20 –5 mm	DP solitary
Hedera	–10 –1 mm+	DP clusters
Ilex	–8 –3 mm	DP long radial files
Leguminosae	–7 –1 mm	DP dendritic
Morus	–7 –60	ring porous
Prunus spinosa *Prunus domestica*	–7 –50+	DP solitary, files and clusters
Quercus	–1 mm –5 mm	ring porous
Rosa	–15 –10 mm	ring porous
Rubus	–12 –10 mm	DP solitary, files and clusters
Vitis	–20 –2 mm	ring porous

Table 3.12 Aggregate rays present (TS and TLS)

Taxon	Aggregate rays present (TS and TLS)	Vessel arrangement (TS)
Alnus	present	DP long radial files
Corylus	present	DP long radial files
Carpinus	present	DP long radial files
Betulu	very rare	DP solitary/files and clusters

Table 3.13 Sheath cells present (TLS)

Taxon	Sheath cells present (TLS)	Vessel arrangement (TS)
Berberis	(present)	ring porous
Clematis	present	ring porous
Cornus	(present)	DP solitary
Leguminosae	present	DP dendritic
Lonicera	present	DP solitary
Morus	(present)	ring porous
Rosa	present	ring porous
Rubus	present	DP solitary, files and clusters
Sambucus	present	DP clusters
Viburnum	present	DP solitary
Vitis	(present)	ring porous

Table 3.14 Undulating ring boundary (TS)

Taxon	Undulating ring boundary	Vessel arrangement (TS)
Alnus	present (at aggregate ray)	DP long radial files
Berberis	present	ring porous
Carpinus	(present)	DP long radial files
Clematis	present	ring porous
Cornus		DP solitary
Corylus	present	DP long radial files
Fagus	present	DP solitary
Hedera	present	DP clusters
Rubus	present	DP solitary, files and clusters
Vitis	present	ring porous

Table 3.15 Rays distended at ring boundary (TS)

Taxon	Rays distended at ring boundary	Vessel arrangement (TS)
Acer	(present)	DP solitary, files and clusters
Berberis	present	ring porous
Fagus	present	DP solitary
Frangula	present	DP solitary, files and clusters
Hedera	present	DP clusters
Rosa	(present)	ring porous
Tilia	present	DP solitary, files and clusters
Ulmus	present	ring porous

Table 3.16 Spiral thickening present/absent, sorted by species (LS)

Taxon	Spiral thickening	Vessel arrangement (TS)
Acer campestre, Acer opulus, Acer monspessulanum	present	DP solitary/files and clusters
Acer pseudoplatanis, A. platanoides	present	DP solitary/files and clusters
Alnus glutinosa, A. incana	absent	DP long radial files
Arbuteus unedo	present	DP solitary/files and clusters
Berberis vulgaris	present	ring porous
Betula pendula, B. pubescens	absent	DP solitary/files and clusters
Buxus sempervirens	absent	DP solitary
Calluna vulgaris	absent	DP solitary
Carpinus betulus	absent	DP long radial files
Castanea sativa	absent	ring porous
Clematis vitalba	present	ring porous
Cornus mas	absent	DP solitary
Cornus sanguinea	absent	DP solitary
Corylus avellana	present – fine	DP long radial files
Euonymus europaeus	present	DP solitary
Fagus sylvatica	absent	DP solitary
Ficus carica	present	DP solitary/files and clusters
Frangula alnus	present – fine	DP solitary/files and clusters
Fraxinus excelsior	absent	ring porous
Hedera helix	absent	DP clusters
Hippophae rhamnoides	present	ring porous
Ilex aquifolium	present	DP long radial files
Juglans regia	absent	DP solitary/files and clusters
Leguminosae	present	DP dendritic
Ligustrum vulgare	present	DP solitary
Lonicera periclymenum	present – fine	DP solitary
Lonicera xylosteum L.nigra	present – fine	DP solitary
Maloideae	present – fine	DP solitary
Morus nigra	present	ring porous
Populus spp.	absent	DP solitary/files and clusters
Prunus avium	present	DP solitary/files and clusters
Prunus padus	present	DP solitary/files and clusters
Prunus spinosa, P. domestica	present	DP solitary/files and clusters

Taxon	Spiral thickening	Vessel arrangement (TS)
Quercus petraea, Q. robur	absent	ring porous
Rhamnus catharticus	present – prominent	DP dendritic
Rhamnus saxatalis	present – prominent	DP dendritic
Rosa spp.	present – fine	ring porous
Rubus fruticosus agg., *Rubus idaeus*	absent	DP solitary/files and clusters
Sambucus nigra, S. racemosa	absent	DP clusters
Salix spp.	absent	DP solitary/files and clusters
Suaeda vera	present	included phloem
Taxus baccata	present – prominent	Gymnosperms
Tilia cordata, T. platyphyllos	present – prominent	DP solitary/files and clusters
Ulmus glabra, Ulmus minor sl., *Ulmus laevis*	present	ring porous
Viburnum lantana	present	DP solitary
Viburnum opulus	absent	DP solitary
Vitis vinifera	absent	ring porous

Table 3.17 Spiral thickening present/absent, sorted by spiral thickening (LS)

Taxon	Spiral thickening	Vessel arrangement (TS)
Rhamnus catharticus	present – prominent	DP dendritic
Rhamnus saxatalis	present – prominent	DP dendritic
Tilia cordata, T. platyphyllos	present – prominent	DP solitary/files and clusters
Taxus baccata	present – prominent	Gymnosperms
Leguminosae	present	DP dendritic
Ilex aquifolium	present	DP long radial files
Euonymus europaeus	present	DP solitary
Ligustrum vulgare	present	DP solitary
Viburnum lantana	present	DP solitary
Ficus carica	present	DP solitary/files and clusters
Prunus avium	present	DP solitary/files and clusters
Prunus padus	present	DP solitary/files and clusters
Prunus spinosa, P. domestica	present	DP solitary/files and clusters

Taxon	Spiral thickening	Vessel arrangement (TS)
Acer campestre, Acer opulus, Acer monspessulanum	present	DP solitary/files and clusters
Acer pseudoplatanis, A. platanoides	present	DP solitary/files and clusters
Arbuteus unedo	present	DP solitary/files and clusters
Suaeda vera	present	included phloem
Berberis vulgaris	present	ring porous
Clematis vitalba	present	ring porous
Hippophae rhamnoides	present	ring porous
Morus nigra	present	ring porous
Ulmus glabra, Ulmus minor sl., *Ulmus laevis*	present	ring porous
Corylus avellana	present – fine	DP long radial files
Maloideae	present – fine	DP solitary
Lonicera periclymenum	present – fine	DP solitary
Lonicera xylosteum, L.nigra	present – fine	DP solitary
Frangula alnus	present – fine	DP solitary/files and clusters
Rosa spp.	present – fine	ring porous
Hedera helix	absent	DP clusters
Sambucus nigra, S. racemosa	absent	DP clusters
Alnus glutinosa, incana	absent	DP long radial files
Carpinus betulus	absent	DP long radial files
Buxus sempervirens	absent	DP solitary
Calluna vulgaris	absent	DP solitary
Cornus mas	absent	DP solitary
Cornus sanguinea	absent	DP solitary
Fagus sylvatica	absent	DP solitary
Viburnum opulus	absent	DP solitary
Juglans regia	absent	DP solitary/files and clusters
Populus spp.	absent	DP solitary/files and clusters
Rubus fruticosus agg., *Rubus idaeus*	absent	DP solitary/files and clusters
Salix spp.	absent	DP solitary/files and clusters
Betula pendula, B. pubescens	absent	DP solitary/files and clusters
Castanea sativa	absent	ring porous
Fraxinus excelsior	absent	ring porous
Quercus petraea, Q. robur	absent	ring porous
Vitis vinifera	absent	ring porous

Table 3.18 Spiral thickening present/absent, sorted by TS anatomy (LS)

Taxon	Spiral thickening	Vessel arrangement (TS)
Berberis vulgaris	present	ring porous
Clematis vitalba	present	ring porous
Hippophae rhamnoides	present	ring porous
Morus nigra	present	ring porous
Ulmus glabra, Ulmus minor sl., *Ulmus laevis*	present	ring porous
Rosa spp.	present – fine	ring porous
Castanea sativa	absent	ring porous
Fraxinus excelsior	absent	ring porous
Quercus petraea, Q. robur	absent	ring porous
Vitis vinifera	absent	ring porous
Suaeda vera	present	included phloem
Taxus baccata	present – prominent	Gymnosperms
Tilia cordata, platyphyllos	present – prominent	DP solitary/files and clusters
Ficus carica	present	DP solitary/files and clusters
Prunus avium	present	DP solitary/files and clusters
Prunus padus	present	DP solitary/files and clusters
Prunus spinosa, P. domestica	present	DP solitary/files and clusters
Acer campestre, Acer opulus, Acer monspessulanum	present	DP solitary/files and clusters
Acer pseudoplatanis, A. platanoides	present	DP solitary/files and clusters
Arbuteus unedo	present	DP solitary/files and clusters
Populus spp.	absent	DP solitary/files and clusters
Rubus fruticosus agg, *Rubus idaeus*	absent	DP solitary/files and clusters
Salix spp.	absent	DP solitary/files and clusters
Betula pendula, B. pubescens	absent	DP solitary/files and clusters
Frangula alnus	present – fine	DP solitary/files and clusters
Juglans regia	absent	DP solitary/files and clusters
Euonymus europaeus	present	DP solitary
Ligustrum vulgare	present	DP solitary
Viburnum lantana	present	DP solitary
Maloideae	present – fine	DP solitary
Lonicera periclymenum	present – fine	DP solitary
Lonicera xylosteum, L. nigra	present – fine	DP solitary
Buxus sempervirens	absent	DP solitary

Taxon	Spiral thickening	Vessel arrangement (TS)
Calluna vulgaris	absent	DP solitary
Cornus mas	absent	DP solitary
Cornus sanguinea	absent	DP solitary
Fagus sylvatica	absent	DP solitary
Viburnum opulus	absent	DP solitary
Ilex aquifolium	present	DP long radial files
Corylus avellana	present – fine	DP long radial files
Alnus glutinosa, A. incana	absent	DP long radial files
Carpinus betulus	absent	DP long radial files
Rhamnus catharticus	present – prominent	DP dendritic
Rhamnus saxatalis	present – prominent	DP dendritic
Leguminosae	present	DP dendritic
Hedera helix	absent	DP clusters
Sambucus nigra, S. racemosa	absent	DP clusters

Table 3.19 Habit sorted by species

Taxon	Habit	Vessel arrangement (TS)
Acer campestre, Acer opulus, Acer monspessulanum	tree – medium to large	DP solitary/files and clusters
Acer pseudoplatanis, A. platanoides	tree – medium to large	DP solitary/files and clusters
Alnus glutinosa, A. incana	tree – medium to large	DP long radial files
Arbuteus unedo	tree – small	DP solitary/files and clusters
Berberis vulgaris	tree – small	ring porous
Betula pendula, B. pubescens	tree – medium	DP solitary/files and clusters
Buxus sempervirens	shrub – medium to large (tree – small)	DP solitary
Calluna vulgaris	shrub – medium to large	DP solitary
Carpinus betulus	tree – medium to large	DP long radial files
Castanea sativa	tree – large	ring porous
Clematis vitalba	liana	ring porous
Cornus mas	tree – small	DP solitary
Cornus sanguinea	tree – small	DP solitary

Taxon	Habit	Vessel arrangement (TS)
Corylus avellana	shrub – large	DP long radial files
Euonymus europaeus	tree – small	DP solitary
Fagus sylvatica	tree – large	DP solitary
Ficus carica	tree – small	DP solitary/files and clusters
Frangula alnus	tree – small	DP solitary/files and clusters
Fraxinus excelsior	tree – large	ring porous
Hedera helix	climber	DP clusters
Hippophae rhamnoides	shrub – large	ring porous
Ilex aquifolium	tree – medium	DP long radial files
Juglans regia	tree – medium to large	DP solitary/files and clusters
Leguminosae	shrub – small to large	DP dendritic
Ligustrum vulgare	shrub – medium	DP solitary
Lonicera periclymenum	liana	DP solitary
Lonicera xylosteum, L.nigra	tree – small	DP solitary
Maloideae	tree – small, medium to large	DP solitary
Morus nigra	tree – small	ring porous
Populus spp.	tree – large	DP solitary/files and clusters
Prunus avium	tree – large	DP solitary/files and clusters
Prunus padus	tree – small	DP solitary/files and clusters
Prunus spinosa, P. domestica	tree – small	DP solitary/files and clusters
Quercus petraea, Q. robur	tree – large	ring porous
Rhamnus catharticus	tree – small	DP dendritic
Rhamnus saxatalis	tree – small	DP dendritic
Rosa spp.	liana, shrub, small tree	ring porous
Rubus fruticosus agg., *Rubus idaeus*	liana	DP solitary/files and clusters
Sambucus nigra, S. racemosa	tree – small	DP clusters
Salix spp.	trees – small to large, shrubs – small to large	DP solitary/files and clusters
Suaeda vera	shrub – medium to large	included phloem
Tilia cordata, T. platyphyllos	tree – large	
Ulmus glabra, Ulmus minor sl., *Ulmus laevis*	tree – large	ring porous
Viburnum lantana	shrub – large	DP solitary
Viburnum opulus	shrub – large	DP solitary
Vitis vinifera	climber/liana	ring porous

Table 3.20 Habit sorted by habit

Taxon	Habit	Vessel arrangement (TS)
Hedera helix	climber	DP clusters
Vitis vinifera	climber/liana	ring porous
Clematis vitalba	liana	ring porous
Lonicera periclymenum	liana	DP solitary
Rubus fruticosus agg., *Rubus idaeus*	liana	DP solitary/files and clusters
Rosa spp.	liana, shrub, small tree	ring porous
Corylus avellana	shrub – large	DP long radial files
Hippophae rhamnoides	shrub – large	ring porous
Viburnum lantana	shrub – large	DP solitary
Viburnum opulus	shrub – large	DP solitary
Calluna vulgaris	shrub – medium to large	DP solitary
Suaeda vera	shrub – medium to large	included phloem
Buxus sempervirens	shrub – medium to large (tree – small)	DP solitary
Ligustrum vulgare	shrub – medium	DP solitary
Leguminosae	shrub – small, medium or large	DP dendritic
Castanea sativa	tree – large	ring porous
Fagus sylvatica	tree – large	DP solitary
Fraxinus excelsior	tree – large	ring porous
Populus spp.	tree – large	DP solitary/files and clusters
Prunus avium	tree – large	DP solitary/files and clusters
Quercus petraea, Q. robur	tree – large	ring porous
Tilia cordata, T. platyphyllos	tree – large	
Ulmus glabra, Ulmus minor sl., *Ulmus laevis*	tree – large	ring porous
Betula pendula, pubescens	tree – medium to large	DP solitary/files and clusters
Acer campestre, Acer opulus, Acer monspessulanum	tree – medium to large	DP solitary/files and clusters
Acer pseudoplatanis, A. platanoides	tree – medium to large	DP solitary/files and clusters
Alnus glutinosa, A. incana	tree – medium to large	DP long radial files
Carpinus betulus	tree – medium to large	DP long radial files
Juglans regia	tree – medium to large	DP solitary/files and clusters
Ilex aquifolium	tree – medium	DP long radial files

Taxon	Habit	Vessel arrangement (TS)
Arbuteus unedo	tree – small	DP solitary/files and clusters
Berberis vulgaris	tree – small	ring porous
Cornus mas	tree – small	DP solitary
Cornus sanguinea	tree – small	DP solitary
Euonymus europaeus	tree – small	DP solitary
Ficus carica	tree – small	DP solitary/files and clusters
Frangula alnus	tree – small	DP solitary/files and clusters
Lonicera xylosteum, L.nigra	tree – small	DP solitary
Morus nigra	tree – small	ring porous
Prunus padus	tree – small	DP solitary/files and clusters
Prunus spinosa, P. domestica	tree – small	DP solitary/files and clusters
Rhamnus catharticus	tree – small	DP dendritic
Rhamnus saxatalis	tree – small	DP dendritic
Sambucus nigra, S. racemosa	tree – small	DP clusters
Maloideae	tree – small, medium to large	DP solitary
Salix spp.	trees – small to large, shrubs – small to large	DP solitary/files and clusters

Table 3.21 Habit sorted by TS anatomy

Taxon	Habit	Vessel arrangement (TS)
Vitis vinifera	climber/liana	ring porous
Clematis vitalba	liana	ring porous
Rosa spp.	liana, shrub, small tree	ring porous
Hippophae rhamnoides	shrub – large	ring porous
Castanea sativa	tree – large	ring porous
Fraxinus excelsior	tree – large	ring porous
Quercus petraea, Q. robur	tree – large	ring porous
Ulmus glabra, Ulmus minor sl., *Ulmus laevis*	tree – large	ring porous
Berberis vulgaris	tree – small	ring porous
Morus nigra	tree – small	ring porous
Suaeda vera	shrub – medium to large	included phloem
Rubus fruticosus agg., *Rubus idaeus*	liana	DP solitary/files and clusters

Taxon	Habit	Vessel arrangement (TS)
Populus spp.	tree – large	DP solitary/files and clusters
Prunus avium	tree – large	DP solitary/files and clusters
Betula pendula, B. pubescens	tree – medium	DP solitary/files and clusters
Acer campestre, Acer opulus, Acer monspessulanum	tree – medium to large	DP solitary/files and clusters
Acer pseudoplatanis, A. platanoides	tree – medium to large	DP solitary/files and clusters
Juglans regia	tree – medium to large	DP solitary/files and clusters
Arbuteus unedo	tree – small	DP solitary/files and clusters
Ficus carica	tree – small	DP solitary/files and clusters
Frangula alnus	tree – small	DP solitary/files and clusters
Prunus padus	tree – small	DP solitary/files and clusters
Prunus spinosa, P. domestica	tree – small	DP solitary/files and clusters
Salix spp.	trees – small to large, shrubs – small to large	DP solitary/files and clusters
Lonicera periclymenum	liana	DP solitary
Viburnum lantana	shrub – large	DP solitary
Viburnum opulus	shrub – large	DP solitary
Ligustrum vulgare	shrub – medium	DP solitary
Calluna vulgaris	shrub – medium to large	DP solitary
Buxus sempervirens	shrub – medium to large (tree – small)	DP solitary
Fagus sylvatica	tree – large	DP solitary
Cornus mas	tree – small	DP solitary
Cornus sanguinea	tree – small	DP solitary
Euonymus europaeus	tree – small	DP solitary
Lonicera xylosteum, L. nigra	tree – small	DP solitary
Maloideae	tree – small, medium to large	DP solitary
Corylus avellana	shrub – large	DP long radial files
Ilex aquifolium	tree – medium	DP long radial files
Alnus glutinosa, A. incana	tree – medium to large	DP long radial files
Carpinus betulus	tree – medium to large	DP long radial files
Leguminosae	shrub – small to large	DP dendritic
Rhamnus catharticus	tree – small	DP dendritic
Rhamnus saxatalis	tree – small	DP dendritic
Hedera helix	climber	DP clusters
Sambucus nigra, S. racemosa	tree – small	DP clusters
Tilia cordata, T. platyphyllos	tree – large	

For Product Safety Concerns and Information please contact our EU
representative GPSR@taylorandfrancis.com
Taylor & Francis Verlag GmbH, Kaufingerstraße 24, 80331 München, Germany

www.ingramcontent.com/pod-product-compliance
Ingram Content Group UK Ltd.
Pitfield, Milton Keynes, MK11 3LW, UK
UKHW050755180926
475701UK00007B/1069

* 9 7 8 0 3 6 7 6 0 5 6 3 6 *